Biosemiotics

Volume 14

Aims and Scope of the Series

Combining research approaches from biology, semiotics, philosophy and linguistics, the field of biosemiotics studies semiotic processes as they occur in and among living systems. This has important implications and applications for issues ranging from natural selection to animal behaviour and human psychology, leaving biosemiotics at the cutting edge of the research on the fundamentals of life.

The Springer book series *Biosemiotics* draws together contributions from leading scholars in international biosemiotics, producing an unparalleled series that will appeal to all those interested in the origins and evolution of life, including molecular and evolutionary biologists, ecologists, anthropologists, psychologists, philosophers and historians of science, linguists, semioticians and researchers in artificial life, information theory and communication technology.

More information about this series at http://www.springer.com/series/7710

Andreas Weber

Biopoetics

Towards an Existential Ecology

Springer

Andreas Weber
und.Institut für Kunst
Kultur und Zukunftsfähigkeit e.V.
Berlin, Germany

ISSN 1875-4651 ISSN 1875-466X (electronic)
Biosemiotics
ISBN 978-94-024-0830-0 ISBN 978-94-024-0832-4 (eBook)
DOI 10.1007/978-94-024-0832-4

Library of Congress Control Number: 2016948232

Printed on acid-free paper

This Springer imprint is published by Springer Nature
The registered company is Springer Science+Business Media B.V. Dordrecht

Feelings are older than humans.
Gavin Van Horn (2017a)

Le poète n'a pas peur de l'imprédictible.
Edouard Glissant (1996:126)

Es ist nichts außer uns, was nicht zugleich in uns wäre.
Goethe (1998)

The unfathomable variety
of yellows and browns and oranges,
cadmium, cream, and crimson,
each hue a singular desire
and a singular lifeline,
a detached focus
on ever the same being.

I feel transpierced with lives,
lathered and rinsed by
vanilla and chocolate wavelengths
caressing my skin and my self-image.
I feel elevated by light, untethered
by the sun, a billion-fold refracted,
answering itself and smiling back in pieces
from the golden days but shortly past.

Brightness in breath,
delicate gravity,
unhinging previous glories without a single
breeze.
Of all the tasty shades so generously
shattered,
of all the dappled things entangled
with each other,
which is the plant's delicious flesh?
Which is a sunbeam's mere reflection?
Which is a dreaming echo in the feast of
ageing
pigments recollecting May?

Stroked by nothing but light.
No mirror's reflection,
but the invitation to partake,
to join the mutual exploring.
Gently smiled at by every singular color's
incandescent iris.

Greeted all with wonder,
yet softly exiled to remember
by the light-footed glances burning
and fading.
Surrendering in uncensored delight
and acquiescence,
as might a roe deer's eyes
before her final breath.

A.W.

Acknowledgments

Books want moments of inspirative fertilization, long stretches of endurance, and endless support of others. This work has been conceived in a brief moment but has then needed exceeding patience and generosity – not only my own, but even more so that of others. I am grateful to my friend Kalevi Kull that he provided me with both. Marcello Barbieri happily embraced the first spark and accepted the then very raw idea into the book series he supervised. My students at Leuphana University, Lüneburg, and at Berlin University of the Fine Arts have helped me to glimpse the flickers of new vision over and over again. Their joy in helping one another in mutual understanding has made me confident.

But the last moment, or rather two-week period, of tinkling inspiration and dogged work, without which the book could not have been brought to a good end, is due to Peter Barnes', Susan Tillett's, and June McCrory's kind support and the welcoming writing shed of Mesa Refuge, Point Reyes, CA. The ocean which is the outside of our inside was also the outside to my eyes as I explored my inside for what had not yet been put into words. I am deeply grateful to my friend David Bollier that he scouted the track leading me to Mesa and to Stuart Kauffman and Katherine Peil Kauffman for so much well-meaning confidence and warm support.

As a living being is not an island, my thoughts are only transformations in an ocean of a thinking and feeling process in which we all participate. Hildegard Kurt, Francesca Michelini, David Abram, Claire Petitmengin, and Natalie Knapp have been sharing most closely my love for the inside shining through the outsides of living bodies. Matthias Mettner made me discover the graceful poetry of Wisława Szymborska in the last minutes before closing time, when it was most needed. And Fernanda Alfieri showed me how to be watched by the moon.

"Even in what we don't know there lies a resemblance" – these lines from Szymborska's poem "The Silence of Plants" could also describe the calm and joyful sharing of the Mesa house and its horizons with my fellow writers Lauren Markham and Gavin Van Horn. Thanks are due for your graceful company, inside and outside.

Contents

Chapter 1
Poiesis

> *"Our scientific theories have failed to explain what matters most to us: the place of meaning, purpose, and value in the physical world."*
>
> Terrence Deacon (2012:22)

In 1976, the Italian writer Italo Calvino published a famous collection of poetic prose, *t zero*. The story "Blood, Sea" recounts a sequence of events narrated from the first-person perspective of a blood cell, alternating with a story about human protagonists, told in the conventional perspective of the third person. In the story, among a lot of other astonishing relations, Calvino explores the fact that the water of the earth's oceans shows a mineral composition which strikingly resembles that of our body fluids.

The blood plasma is the sea in which life once began. This ocean still fills us, as it fills all other lifeforms. Calvino imagined a narrative told by a blood cell, a cell, which is suspended in this primordial ocean within our bodies. He told a story from the perspective of life itself, or rather from the perspective of the life-giving ability of the primal fluid and its invitation to make intimate connections. He spoke from the standpoint of an outside which is also an inside.

Calvino invented "Biopoetics" *avant la lettre*. He envisioned a first-person-account of what is not human through our shared qualities, through our participation in a vast web of transformations. For Calvino, the poet, it was only evident that we are able to make statements about this network of changes and exchanges because we are a part of it, and we are concerned by it, as we are by our own fate. Poetic creativity is the power to know something through intimate participation. It is the power to know through recognizing which familiar qualities lie in what is totally different, and to find unexpected strangeness in oneself.

When we walk the seashore, the fringe of the "true" ocean, which lies before us and flattens the beach with the incessant caresses of its liquid weight, we involuntarily repeat Calvino's insight. Humans are drawn to the ocean's shore, or to a journey across its waves, only to despair at accounting for the character of this yearning, this desire for closeness pulling them to search contact with what is very different to our outwardly dry and tight bodies, but which is nonetheless their central mystery. Why do we love the sea so much? Why are humans so inevitably drawn to what in many respects seems to be their anthithesis, blunt contradiction to their dependency on firm land, on the "the green, gentle, and most docile earth" (Herman Melville)?

A. Weber, *Biopoetics*, Biosemiotics 14, DOI 10.1007/978-94-024-0832-4_1

The sparkling roar which greets the long curves of the pacific shoreline at Point Reyes, north of San Francisco, shining in a blinding white which seems to concentrate the sun's entire energy in the crest of the breaking waves; the cosy algal thickets at the underside of the boulders which are strewn into the trembling froth; the cliffs and rocks steaming with wetness and salt dust against an oblique western sun; the squeaking seagulls, the tiny ducks who relish in their swift dives just when a breaker's crest is about to topple down on them – all these bodies and lifelines are present before our eyes and beyond our skins. But for the secret magic of perception of someone who has been born as a part of the world, and who needs its stuff to still her hunger and its liquids to quench her thirst, the rolling waves crashing on the shore are not outside, but inside.

"Blood, Sea": the deeper message of Calvino's work is that the ocean is not an outer object, but truly our inside. It is our inside that happens to be able to be touched and smelled, and which can drown us. Here, although not many might have realized it at the time of publication of *t zero*, Calvino is extremely radical. He is no psychologist, nor a scientist, and at the very least he is a cultural criticist catching the weak human imagination projecting details of its cultural heritage to the outer world. Calvino is a poet, and as such he knows about the fact that true novelty in this world, and also true experiences of connection, only arise through the exchange, the breakdown and recreation of what is real.

He understands that our body is the way the sea thinks itself through being folded up into a confined space. He understands that our body is the way the ocean imagines itself having an outside. He knows that the ocean is the way we can experience our inside – not as a symbol, but as an existential reality. The experience of the ocean is a way my body can think its inside to which it otherwise has no access.

In "Blood, Sea", the ocean is not a *metaphor* for inwardenss. It *is* inwardness, seen from the perspective of a body, who is always both, metabolism, proteins and bones, but also concern, feeling and lust. If we connect to the ocean, Calvino suggests, we experience ourselves, but we experience that in us, which is so deeply inside that it is completely unknown. Our inner dimensions are oceanic. When we experience ourselves, in truth we have an experience of the ocean, of its vastness, wildness, bleakness, of its fecundity, desertion, and its unfathomable life-giving potential.

Our body is the fashion in which the ocean thinks its own enfolding into a tiny space. The ocean is the manner in which my body experiences its own inside. When I connect to the ocean, I can make contact to this inside, which is always there, as it is simply the means to make any experiences, but which I still cannot really see, as it is the means through which I am able to see in the first place.

The ocean which carries me on its waves and makes me feel carried, and embraced, and chilled, and rocked, and which makes me taste my salty self on my own lips, is an endless variation of how it is to be, not as a mere body in space, but as a part of that all-encompassing space which is me and the ocean. Floating in sea water means experiencing my inside by being touched by that which is inside and which at the same time totally is the other. It means experiencing myself as ocean by letting meet my inside and the other who is also an inside.

The ocean reveals itself as inwardness, as I connect to it through my body, and experience inwardness, and, in the roar of the waves, and the salty taste of the spray, this inwardness shows itself as being matter, and form. It shows that every being is the constant transformation of one into the other, of inwardness into matter, and of matter into experience. It shows that in experiencing, we are outside, and that we, as bodies with an outside, unfold meaning and existential concern. We see with our bodies. Through the touch of something which is a body outside of our bodies we can see ourselves, not as an outside, but as an inside.

Being an Organism Is Desire

The hypothesis of this book is that Calvino in his idea did not follow only a beautiful, but somewhat idiosyncratic quirk of a poet, but that he rather explained a basal truth. This truth is that reality – and in it explicitly the biosphere – is alive. This aliveness means that it not only shows a physical surface, a causal-mechanic behavior in space and time, but that it is constantly revealing meaning and inwardness. Biology is the sphere of existential relations expressing this inwardness, hence, it is poetic.

The idea which Calvino unfolds is a first-person-analysis of how organisms connect with one another. It is absolutely accurate, particularly because it does not only recount the story of a rupture and a separation, but also an intensification, a mirroring in the play of transformations: the general poiesis of reality translated into the poetics of words.

The biosphere is deeply poetic, not only as a source for romantic storytelling or spiritual nature writing, but as a means and matter of its physical functioning. Being an organism is about constant transformation, about developing a standpoint of concern, about desiring to be a self which yearns for connection with other selves (and this also way below the human needs, as mating, feeding, shelter concern every being), and which takes up chunks of the world as food only to incorporate them into the own physical body.

Only if we understand the poetic dimension of biology we understand ourselves, and reality. What is missing in the description of the world as sum of lifeless processes which for long has dominated our view of reality is the existential reality of inwardness, concern and expression practiced by living bodies. This lack accounts for the strange situation that humanity at present knows incredibly many technical details about life, but at the same time has never been more destructive towards other lives of other species.

We need to understand life as a not only material, but deeply sense-creating phenomenon. Biology is the realm of meaning, starting from the meaning of DNA bases, and unfolding into the behavioral signal cues, facial expressions, musical scores, and biopoetic books. I therefore subscribe to the biosemiotic turn in life sciences (Uexküll 1980; Kull 1999; Hoffmeyer 1997, Weber 2003, Weber 2010a, b; 2016).

I think it is important to understand biochemical processes and behavioural repertoires from a semiotic standpoint, as meaning generation and interpretation. But we still need to explicitly highlight that meaning generation truly means to feel, and hence that organisms live in existential, felt, highly biased realities which they experience through their selves. We need to see that life is an embodied semiotic process, operating on meaning, not only on cause and effect, and that the underlying semiotic processes have a side of inwardness which permeates all biological life, and which we continuously experience as our subjective existence (Weber and Varela 2002; Weber 2016).

These inward experiences are existential, not epiphenomenal, because living beings are subjects concerned with their survival and hence with what is good or bad ("yum" or "yuck", as Stuart Kauffman (2002) puts it). Meaning generation has a more than technical side. It is about experienced interiority, about true concern, about an absoluteness of the individual's own existence, which we cannot strip from the picture unless we want to make it inaccurate. It is something that we organisms know from the inside. This knowing must complement our view of what it is like to be alive. The biological world is the deep experience of existential values. It is the experience of feeling – as inwardness, but also as expressivity of the other insides of the endless number of bodies which constitute the living world. Because this living world is an embodied expression of inwardness, we can call it poetic.

Sensuous Selves

This book tries to understand the still enigmatic and much contested reality of this poetic dimension from a novel account of the construction of an embodied biological identity. For this, it follows the explorations of biosemiotics, embodied cognitive science, and existential biophilosophy. From here, I want to develop a different understanding of the dimension of lived experience, which includes body, hence physiology, and perception, hence inwardness.

In this book I propose to understand the current paradigm shift in biology as the origination of a biology of subjects. A description of living beings as experiencing selves has the potential to transform the current mechanistic approach of biology into an embodied-poetic one, culminating in a poetics of nature. We are at the right moment for that: The findings of complex systems research, autopoiesis theory, and evolutionary developmental biology are converging into a picture where life can not longer be described in terms of causal mechanisms. Instead, organisms bring forth themselves physically and thereby generate a poetic standpoint. They interpret external and internal stimuli interfering with their self-creation according to embodied values. This can be observed empirically during embryonic development, where genetic instructions do not act as orders, but rather as weak interferences being interpreted by a self-maintaining developmental centre which foremost strives to maintain its own integrity.

Organic subjectivity opens the living world to a poetic perspective. Since any encounter has a meaning and is interpreted accordingly, it creates a self. This self experiences external and internal stimuli as values. It is coextensive with the material dimensions of biochemical processes, but it is their other, or symbolic, side. By this process the subjective perspective of organisms is open to other's experience. Meaning and value become visible, as they are generated in material, embodied form. Instead of being separate from nature as pure "mind" or "language", we all share with any other being the same "conditio vitae" of experienced meaning and expressive feeling.

If organisms cannot be described exhaustively in terms of mechanical laws, we must redraw the border between empirical science and interpretative science. By subjective interpretation of encounters according to needs, living beings seem to follow the paradigm of meaning and interpretation, which had been reserved for the human sphere alone. Living subjectivity might even be turn out to be a primary poetics – the first step in the way through which the world experiences itself. Interpretation from this viewpoint is no longer based on human language or texts alone. Rather, interpretation starts with organism. Subjectivity, not objective relationships, are at the forefront of any biological experience – including our own.

The poetic approach accepts that the poetic, feeling and expressive subjectivity of organisms is a physical factor. It is not an epiphenomenon but rather its opposite: the foundation from which an explanation of life has to start. The poetic approach discovers meaning, value and feeling at the centre of a physics of organism. It takes it not only as one of many possible viewpoints, but as a necessary element of a scientific description of life. The endeavour of the following pages is to explore, in a preliminary sketch, how that biological poetics, the new science of ubiquitous *poiesis*, could reasonably be laid out.

Biopoetics as Paradigm for Living Relationships

Poiesis, the perception of the world through inwardness and the expression of this inwardness through transforming the individual relationships with this world, is not alien to a biological description of organism. In order to see this connection, however, we have to fully embrace the current changes in life sciences. We need to transform biology from the objective description of meaning-free mechanisms and turn it into what it in truth is: a partaking in a process of deepening connections with other life.

This change is happening everywhere in the science of biology: We only need to become aware of it. Biology is undergoing a profound reassessment of its core premises, much as physics was transformed when relativity and quantum theory were discovered about a century ago. Already two decades ago the molecular biologist Richard Strohmann (1997) foresaw a paradigm shift that he named the "organic turn in biology". By 2016 many of his assumptions have been empirically confirmed. The theoretical foundations of the classical molecular-evolutionary model in biology have now been called into question. "Weak linkage" instead of causal mechanics defines evolutionary developmental sciences (Kirschner and Gerhart 2005). Epigenesis has come back. Epigenetic regulation plays a much more important role than previously thought, which means that individual organisms can influence the fate of its own genes (Jablonka and Lamb 2005). It is now well-established that parental experiences can be passed on genetically (Bauer 2008) and even that how a child is treated individually may directly influence the offspring's genome (Powell 2009:8). The emerging paradigm of regulation even questions the concept of the biological species: the majority of organisms must be viewed as "metabiomes" consisting of thousands of symbiotic, mostly bacterial taxa (Ley et al. 2008).

We have become aware that an organism can be regarded as an ecosystem or a "super-organism" built from innumerable cellular "selves" – and that a given organism is not simply the result of a linear cascade of causes and subsequent effects. Developmental genetics, proteomics and systems biology are beginning to appreciate self-production and autopoiesis as central features of living beings. Autopoeiesis, literally "self-creation", is a term introduced by Chilean biologists Humberto Maturana and Francisco Varela (1980) to describe the capacity of an organism to continuously generate and specify its own organization autonomously. Genetic coding, developmental and regulatory processes are increasingly discussed in terms of an organism's capacity to interpret and experience biological meaning and subjectivity (Kirschner and Gerhart 2005).

These findings not only challenge the standard empirical approach to organisms. They transform our underlying assumptions about what life is. Is an organism a machine, assembled from parts that have to be viewed as still smaller machines or sub-assemblies? Or is life a phenomenon in which subjectivity, interpretation and existential need are key forces that cannot be excluded from the picture without distorting our understanding of how an organism functions and without obstructing the path to further explanations?

In the emerging new picture, organisms are no longer viewed as genetic machines, but basically as materially embodied processes that bring forth themselves (Weber and Varela 2002; Weber 2010a, b). Each single cell is a "process of creation of an identity" (Varela 1997). The most simple organism must be understood as a material system displaying the intention to maintain itself intact, to grow, to unfold, and to make a fuller scope of life for itself. A cell is a process that produces the components necessary to maintain their ongoing production – while the materials, carbon, nitrogen, oxygen, phosphorus, silicon flow through it. Biology is discovering that sentience and felt expression in organisms are not just epiphenomena. Scientists like the Harvard embryologists Marc Kirschner and John Gerhart (2005), the Copenhagen and Tartu-based theoretical biologists Jesper Hoffmeyer (1997, 2014a, b) and Kalevi Kull (1999, 2015a, b), the Berkeley based biologist Terrence Deacon (2012) the and science theoretician Elizabeth Fox Keller (2002) are starting to acknowledge that meaning and expressiveness are deeply rooted in the heart of nature.

Eminent biological and systems thinkers as Lynn Margulis (1999), Francisco Varela et al. (1991), Alicia Juarrero (1999), Stuart Kauffman (1996) and Gregory Bateson (1972), Ezequiel Di Paolo (2005) and Evan Thompson (2007) have begun to draw a picture in which organisms are not longer machines competing with other machines, but rather a natural phenomenon that "creates" and develops itself in a material way while continuously making and expressing experiences. Being alive, these researchers wish to show, is not a case of cause-and-effect alone, but also a complicated interplay of embodied interest and hence feelings. Brain researchers like Antonio Damasio (2000) have shown that emotions, not abstract cognition, are the stuff of the mind.

If we consider all these changes in contemporary biology, a completely different picture of the living world necessarily emerges. We are starting to see that humans do not exist at the exterior or edge of "nature," but are deeply interwoven into the material, mental and emotional exchange processes that all of the more-than-human world participates in. This is leading biological sciences to a major paradigm change of the sort that physics experienced a century ago.

Physical science for a long time has shown that the separation of observer (subject) and observed phenomenon (object) is an artifact of causal-mechanic, linear thought. In quantum physics, there is no locality or temporal chronology. Rather, any event can be connected to any other. The physicist David Bohm (1985, 1990) has called this the "implicate order" of the cosmos. This view not only calls into question locality and chronology, it blurs the separation of physical and psychological reality. We exist in a space-time that is a continuum of "insides" (meanings) and "outsides" (bodies).

"Enlivenment": A Program for a Poetic First Person Ecology

The task of this book is to build a unifying frame for the emerging new biology of feeling subjects. In previous works I have called that endeavour "enlivenment", as it is meant to reconsider aliveness as a basic reality and as an expressive nexus between insides and outside (Weber 2013, 2014, 2016). Reintroducing felt aliveness in our picture of the world means to deeply challenge the predominating worldview even beyond the already radical changes in biosemiotics and cognitive biology. It basically means to come to terms with the enlightenment separation into rational controlling and controllable matter. This new enlightenment, an enlightenment 2.0, I name "enlivenment" (Weber 2013; Weber and Kurt 2015).

I am non alone in my call for a fundamental ontological shift regarding organisms. Already Wilson (2012) has called for a "second enlightenment" which reconsiders a basic dimension of freedom and meaning in nature. Wilson elaborates on the hypothesis that man is under constant tension between the needs of individual and group selection. *Homo sapiens*, argues Wilson, is biologically determined to freedom of choice. This determined indeterminacy, so the argument, gives rise to culture and the realm of meaning in the human semiosphere.

The pledge for a "second enlightenment" and an "enlivenment" wants to open up the heart of biological thinking, and break with the strongly deterministic notion of the "egoistic gene" and related concepts. We should take this idea seriously. It is not much different from using the biosemiotic view of life as meaning as a concept that can sketch a poetics extending from lived body to the human semiosphere. The "enlivenment" is a reassessment of aliveness as a dimension of existential relation beyond the old dualistic domains (Weber 2013, 2016).

To challenge "enlightenment" with a new approach means to put life-as-meaning center stage. If we incorporate the poetic dimension of the living body, we are working on a general theory of aliveness and therefore on an existential ecology of being-in-the-world. It builds upon the following propositions (Weber 2013):

- Natural history is not the functional interaction of organic machinery, but the embodied unfolding of freedom, autonomy and agency.
- Reality is alive: It is full of subjective experience and feeling; organic feeling is the prerequisite of any experience and of any rationality.
- The biosphere is a material and semiotic interrelation of selves.
- Embodied selves come to be only through others: The biosphere depends on cooperation and interbeing.
- The biosphere is paradoxically cooperative: Its relationships are unfolding out of antagonistic, or incompatible processes: matter/form, code/soma, ego/other: incompatibility is needed to achieve life in the first place.
- The individual can only exist if the whole exists and the whole can only exist if individuals are allowed to exist.
- The experience of being alive, of being in full life, of being in the joy of full life is a fundamental component of reality: the desire for living / experience to become one's own full self is a general rule of biological worldmaking, both interior/experiential and exterior/material.
- Death is a reality, it is inevitable and even necessary as the precondition to allow for the individual's striving to keep intact and to grow. Death is an integral component of life.
- The living process is open. Although there are general rules for embodied identity in interbeing – aka accomplished life–, its form and way is entirely subject to openness.
- There is no neutral information, no general ("scientific") objectivity, but a common experiential level of understanding, interbeing and communion of a shared "conditio vitae". New structures and levels of enlivenment can be made possible through living imagination.

From these structural observations, it seems possible to complete the existing empirical-ecological worldview (which is basically a technology of exterior material nature viewed as pool of resources) with an interior, or intentional aspect.

Poetic Space: The Scope of First Person Ecology

Following these observations, we can define life as a poetic space, which encompasses both material processes and meaning relations, joining them together to a lived experience, which is "felt" or subjective from the inside and "sensous" or "expressive" from the outside. The poetic space is not inside as "spirit" but inside as body, which can be best understood as metamorphic material which is always meaningful. The individual share of poetic space can be called "core self" (Panksepp 1998).

I want to argue that the phenomenon of the living starts in this "hybrid" region of being neither inside nor outside. The lived space is a continous communication of felt meaning which at the same time is materially embodied and "ideally" poetic. The poetic space of the living hence has to be conceived of as really one space. This idea breaks with any notion of primacy of either matter or symbolic relationships and hence in a radical way is nondualistic. There is no outside to this poetic space, the poetic space encompasses both organic and anorganic matter.

The imaginary process of this poetic space can be subject to transformation from both sides: through material manipulation but also through imaginative creation. The poetic space is the only transformable space that there is. The poetic space of the living is open to new interpretations, new framings of utterances of self-expression and can be physically transformed in that way.

In this respect the poetic space of the living is also the place of freedom, where in a certain range of existential flexibility no preestablished values of good or bad exist. It is evident that material impacts can be of existential importance, but also signs can be of healing or mortal influence. Both levels, however, are never separable and always joint: Because the poetic space is the only space where living occurs, any material impact has "felt" consequences and viceversa.

From this point one could argue that any sense-making processes should take place in this lived poetic space. Also, any damage done to members of the poetic space, predominantly lived beings, reduces its scope and creative power. From this, there are direct implications for our daily dealings with our being alive, the life on this planet and our relationship to the living, which at the moment are treated as rather technical or structural matters. Learning from the idea of poetic space, it should follow that any process of imagining reality has its greatest potential to be alive if it is a poetic – or artistic – process.

In our time the important philosophical discussions – empirical rationality, human freedom as a rational actor, instrumental reason in economics – have been accused as deficient, provoking a mad scramble to salvage them as coherent perspectives. The real issue of our time, then, is to activate a new language. After 300 years of Enlightenment thinking, the challenge is to redefine aliveness and humanity within it by complementing *techné* with the concept of *poiesis. Techné* means explainability, analysis and successful replication. *Poiesis*, by contrast, means creative self-realization – an element that brings forth reality, that cannot be suppressed, and that can never be sufficiently understood to be successfully controlled.

In the end, everything is *techné* in one sense – but in another, everything is also *poiesis*. *Techné* is cause and effect, control, management, understanding, exchange. *Poiesis* is inner goal-directedness, bringing forth oneself, giving oneself over, self-expression, feeling, and accepting. *Techné* is planning and sustainability. *Poiesis* is the "wasteful" promiscuity of creation. Life needs both. Reality is both. Creative transformation grows out of the tension of this contradiction without ever resolving it.

Chapter 2
Subjectivity

"It is said that contradiction is unthinkable; but the fact is that in the pain of a living being it is even an actual existence."

Georg Friedrich Wilhelm Hegel (1969:770)

Nature is not only a generator of form, but also of meaningful worlds. We know this, as we are examples for this process, and as meaning-generation is the way we experience our existence in connection with others. But for a biological approach, meaning is still difficult to integrate into a description of the biosphere, let alone the inevitable company of its experience: feeling, subjectivity, and inwardness.

This problem became exacerbated when physics changed from a purely deterministic science to an open-framed exploration of relationships in the kosmos. Physics – at least in its more advanced theoretical areas – has become the paradigmatic place of subjectivity and connection. Biology, the description of beings capable of subjectivity and connection, for a long time could not participate in this reorientation. By the mechanistic approach of biology organism had become an easily treatable machine. When Newtonian physics crumbled, it opened the field for a new assessment of the problem, without really being able to solve it. Erwin Schrödinger's (1944) famous paper "What is Life?" illustrates this problem in a paradigmatic way: The physicist acknowledges here that "living matter operates in a way that cannot be reduced to mere laws of physic".

At the same time Schrödinger is unwilling to accept a view of the organism that puts it outside of physics. In a way, life had been more easily explainable in a completely mechanistic universe than it was now in an indeterminate one: in a thermodynamical world the regularities of the living become unexplainable. Machines degenerate over time. Organisms, although also physical bodies, unfold and multiply. In regard to the second law of thermodynamics the living world had become a severe theoretical problem.

Schrödinger's paper has become famous for his attempt to solve the enigma encountered: he postulated that organisms obviously had to "feed on negative entropy" to keep their orderly state in a universe where the second law is sovereign and where the laws of physics in general are sufficient for any explanation. The term "negative entropy" has persisted until today as a common enigma in the "folk philosophy" of Biology.

A. Weber, *Biopoetics*, Biosemiotics 14, DOI 10.1007/978-94-024-0832-4_2

The most powerful attempt to answer Schrödinger's question has been undertaken by Francisco Varela. Varela does this by altering the preconditions for our reflection about organism. He substitutes the tacit assumption that a living being must be a passive machine by the idea of something which is actively caring for itself – a subject.

Varela seeks to explain the organism's persistence over time not in terms of exotic physics, but by reintroducing something into the picture which had been buried: the standpoint of a self. Varela attacks the old difficulty of the living organism right by the front. He writes: "I want to start declaring that I think that understanding of organisms and the living *is* possible, that defining these terms in a satisfactory manner is not a utopian dream, and that we even have a good deal of the road already charted. However, this is under a fundamental condition: that the autonomy of the living is highlighted instead of forgotten, as it has been" (Varela 1997:73).

The Construction of Self and World

By this, Varela provides a substantial definition of life. It is simple, but it turns over the logic of non-contradiction, which has been the basis for western science. "Organisms", Varela (1997) writes, "are fundamentally a process of constitution of an identity". Organisms bring forth an identity as a material process: the observable *telos* of metabolism is to perpetuate itself.

A living entity produces itself and all its components autonomously. It is distinguished by the ability to retain its integrity in the face of changes in its environment. It generates the structure as well as the border of its surroundings. The whole organic machinery has one primary goal: it produces exactly the components that have produced it. Ribosomes enable proteins to be formed that become ribosomes. Cell membranes which are brought forth by the cell interior shelter the cell interior to produce membrane components, and so forth.

A cell's activity is to 99 % concerned with the maintenance of itself. This observation can be expanded from single cells to multicellular life forms. A very high degree of self-relatedness is also valid for higher levels of more complex organisms. In vertebrates, the overwhelming bulk of all neuronal activities has nothing to do with external stimuli but is an occupation of self with self (Varela et al. 1991).

On the other hand, this cellular or multicellular-somatic self so obsessively concerned with itself does not have any constant material essence: it is important to see that the living entity exists as a certain self-identical structure in space and time, although it is at no moment materially identical with itself. Matter passes through its changing spatial arrangements. Only the fact of being alive keeps this circuit closed. When an organism dies, the process comes to an end, and the components behave as normal chemical compounds tending to assume the highest possible degree of entropy: they decay.

The organization of that which lives is therefore characterized by the conjunction of two different ontological realms. Unshaped matter and the process of regulation only together make up the proper reality of the organism. This processual circularity is a fact that biology must take into account. The process of living takes place in normal matter, only the latter is organized in such a way that it shows autopoietic behaviour. This behaviour introduces a goal which is not inherent in matter.

The autonomous encounter with reality evaluates everything in reference to to this individual goal and acts accordingly, and therefore is not completely causally determined, Varela calls "cognition". He uses the term not in the classical sense of the cognitive sciences (which understand by "cognition" the logical operation of symbols) but in an attempt to emphasize the creativity of opening up a world in interactions relative to the system. Symbols are created according to the organism's existential situation, when it constructs itself according to the meaning of external and internal stimuli for the ongoing self-production process. In that kind of cognition, organisms create relevance by separating the outside from themselves, while at the same time being dependent on it.

From where does this drive of matter to form autopoietic loops stem? Whence this desire for wholeness? Even though we are not able to give an empirical answer, we cannot exclude the sheer fact that the tendency to autopoietic complexity exists from our picture of life. We have to take this tendency into account as something just normal in the cosmos. There is a bulk of empirical evidence that shows that obsession with complexity – an obsession to enhance life – is everywhere.

Research into self-organization and complexity shows an increasing tendency to creativity, to the unfolding of higher complexity and "order for free" in our world (Kauffman 2002, 2016). We can view chemical autocatalytic networks as precursor systems of the cell's autopoietic behaviour. Autocatalytic networks are made of a high number of different reactive components, just like it is the case for a cell, which contains a host of different molecular classes. At a certain level of complexity these components start to catalyze new reactions. With growing interconnectdedness, there may be as many reactions catalyzed as there are classes of components. This leads to a "crystallization" of structure which starts to sustain itself. (For a detailed discussion see Weber 2016; Kauffman 1996, 2002).

The transition from the physical level of energy to the level of significance happens every time the organism – on the simplest level, a single cell – manifests as a "whole". Meaning therefore is co-extensive with life. It is co-extensive with the closure of a physical self. Every relation of energetic exchange is a relation of signification brought forth by the existential background of the organism. By self-confirmation, the non-self as Umwelt is separated off; and this separation then opens the possibility of interaction with the Umwelt, for better or worse. Seen from the organism's perspective, the organism's domain of action is its unique standpoint. This standpoint separates the own cell-body, which is to be developed and protected, from the surroundings, which are thus valued as good or bad prospects for the self-realizing organism (Weber and Varela 2002).

Organisms Are Not Machines

Every living being interprets the world according to its needs, and its desire to carry on and prosper. The world thus gains, in the same movement by which self and other are divided, existential significance for the emerging self. A "perturbation", in autopoietic terms, is a stimulus that interferes with the organism from the outside, causing it to react in a way that is determined only by its inner processes and states.

When organisms are conceived of as autopoietic systems, meaning is their fundamental dimension of existence. Probably here lies the true boundary that separates the organic realm from pure matter. It is the organism's paradoxical dependence on its surroundings that lends an irrevocably existential value to those surroundings.

Living beings are those entities which are first and foremost concerned with themselves. All other descriptions that overlook this self-relatedness must miss the most important point of defining life and the living. Any comparison to an artificial machine leads the wrong way, as no machine intends its own continuation. Machines, as man has constructed them, process matter, consume energy and produce things. And they generate waste, which is also not found in living nature; here used-up material is always a starting point for new sense-creation by other biota.

Artificial systems do not produce themselves. Even the most advanced self-constructing robot depends on high-entropy building blocks that have been prefabricated by man. Life in contrast to that is the obsession by which a lump of stuff tends to maintain a certain form and a certain way of acting. A living cell is always involved in self-creation. It is the material realization of the principle of subjectivity.

In a manner reminiscent of the picture that quantum theory painted 80 years ago, such a perspective departs from linear causality: cells react with a certain autonomy. How they see the world influences the way this world is. How they act does not only depend on an objective constellation, but also on their inner state, on context and meaning, as well as on what an observer is doing or not doing. Organisms are highly subjective and existentially concerned, not trivial and deterministic. They are certainly not molecular clockworks telecommanded by their genetic software.

This persistent pursuit of the own being and well-being is the basic character of life, from a simple cell onward. The most important feature of a cell is the fact that it consistently restores itself and brings forth all its constituents. Cells show a breathtaking perseverance. In a steady exchange of matter with their surroundings they resemble batteries that continuously recharge themselves. They spiral upward on their genetically inherited molecules, but they do not follow their code as inevitable instruction, but rather in a continuous dialogue between self and other, subject-pole and surroundings, not determined, but orchestrated by their genes.

The major portion of biochemical activity is invested in keeping the inner order alive, to stabilize it, to enlarge it, to rebuild it. An organism maintains itself against the steady pull of disorder, against quantum fluctuations, against the mass of minuscule breakdowns. A single bacterial cell continually repairs DNA errors. Otherwise a cell's life would end within moments. It would gradually decompose.

And just that happens when death occurs, when a living being is no longer able to define itself as a centre of activity. But there is more: life does not content itself with defining itself as a centre. It reaches out to unfold more of itself. Life is longing for further life to subsist: what we perceive as sustainability is always enhancement.

The prototype of any subjectivity thus is a subjectivity of body, not of mind. Its defining character is the autonomy of form over matter while at the same time form is dependent on exactly this matter. The living cell governs the atoms that make it up. The identity which is brought forth through it holds and moulds matter. Therefore we can introduce another term by which we can summarize the qualities of subjectivity and autonomy – qualities that are themselves already rather unexpected in a biological description of organism: an organism displays a certain degree of *freedom* (Jonas 1996).

A bacterial cell has many more possibilities to choose from compared to, say, a grain of sand. A microbe does not ponder about how to meaningfully spend the next day, but neither does it lurk around like lifeless matter. It decides and chooses according to needs which arise from the tenacity with which it realizes itself anew in every instant. It is free because it shows an intention to carry on – or rather: it is free by necessity.

In this new perspective on biological science cells appear as units of will. This cellular desire is not like ours, personal and conscious. But still it is the will of life to unfold itself. Therefore it is something which we also find within ourselves.

Interpretation in Action

The first step towards an embodied poetics is to take into account the intentional habit of matter whose consequences I have described so far. This is indeed what many researchers do at the moment: They start to rediscover intentionality as an empirical fact. Actually, this is no wonder. Although intentionality has 'gone lab' only very recently, it has been lying in wait for years. Already the "old" genetic paradigm was formulated in the cryptosmiotic language of DNA as a code or a text, hence demanding poetic – at least interpretative – understanding, not causal determination (Hoffmeyer 1997). But now a much broader empirical view of embodied interpretation comes into sight.

In the emerging picture, genes do indeed play a central role – but it is a role very different from that which most authors previously assumed. "The miracle of complex life is more amazing, yet ironically simpler, than anyone ever expected", states developmental biologist Sean Carroll (2005). Experiments have shown that many biophysical structures – such as fur patterns or leaf surfaces – reflect processes of self-organization without the involvement of precise genetic instructions. Biological order arises "for free". But in the past few years researchers have observed that whole embryonic growth centres unfold largely independently of genetic commands. Embryonic tissues and organs organize themselves as centres of concern: they keep themselves closed and intact over time (Kirschner et al. 2000; Kulesa and Bonner-Fraser 2000).

The American biomathematician George von Dassow and co-workers (2000) recently found that during somatic development a given number of genetic switches can be connected in only one way to bring forth a specific body trait, e.g. an organ. But if such a network has been started and runs, it is resistant to a whole range of influences. The group simulated *Drosophila's* segment polarity network by computing a set of 136 differential equations symbolizing a large number of parameters as half-lives, diffusion constants, binding rates etc.

"The network's ability to pass our test is intrinsic to its topology rather than to a specific quantitative tuning. There are so many diverse solutions that the notion of a global optimal parameter set [of genetic commands] makes no sense", von Dassow et al. (2000) say about their work on genetic regulation of the chicken developmental centre. Elsewhere they comment that "The simplest model that works at all emerged complete with unexpected robustness to variation in parameters and initial conditions". Once it kicked in, it remained going. Dassow et al. are convinced that "robust gene networks are the only networks natural selection can evolve" – networks which hence are not guided hierarchically but rather form a chorus of independent components (Fox Keller 2002).

DNA is a scaffold for the flesh, not its blueprint (for an enlarged idea of DNA as a "scaffold" see Hoffmeyer 2006). DNA does not carry instructions, but rather allusions: the body must "read" the genes according to its overall capacity to understand them. It must interpret the sequences and transform them into sense. We should no longer view DNA as a machine code to execute orders, but as a kind of musical score in relation to which the cell can choose different instrumentations according to its status. The DNA is much more part of the cell's metabolism than the blueprint metaphor suggested.

Contrary to what had been supposed, then, the genome does not carry a homogenous blueprint for a living being. The different components that assemble the hereditary molecules are better understood as an assembly of scattered fragments and pieces of relative meaning which alone could never make up an organism. Most of the genetic material does not encode proteins, but works as switches in a wider developmental and metabolic network.

The formerly so-called "junk DNA" plays a major role in this cellular "neuronal net". Most of it is involved as genetic switches and toolbox genes in metabolic and developmental regulation, not in gene-product coding. And the genes themselves are not forever fixed in one state. As would be suspected of flexible players in the metabolic game, genetic sequences do not represent permanent law, but can be processed by the soma, e.g. via extended "capping" or by methylation of genetic promoter regions.

Genes should thus be viewed as the "nervous system of the cell" rather than as mechanistic sets of orders for ready made proteins. The genome contains several pathways for genetic switches (e. g. via genetic enhancement or repression and via miRNA) which are wired in a way that reminds of a massively parallel computer (Bauer 2008). DNA is therefore less an information storage and more a highly complicated cognitive subsystem concerned with meaning.

The choice about which genes are activated and actually produce proteins is dependent not on an unequivocal "order" but rather on the outcome of a virtually endless number of possible connections. Not blunt genetic information, but rather the network and its emergent properties – and above all its emerging autonomy – decides which organism will be built up. But the decision is taken only by doing. The computing of the developmental centre *is* the unfolding of the organism. It is enacted in real time when the developmental process has been started. Once switched on, tissue developmental centres follow their trajectories without paying much attention to (non-lethal) disturbance.

The developmental pathway should therefore be compared less to a chemical reaction (as is still the norm in biology) but to complex behaviour. Cell arrays, e.g. in liver tissue, differentiate when they are "tuned" in a certain way by the sum of all different inputs by every other cell and by all additional triggers in their close as well as further vicinity. Like a songbird which starts to build its nest when longer daylight and warm sunrays have raised the levels of hormones in its blood, cells respond to the DNA's message according to their individual situation, each time in a different adapted way.

Kirschner and Gerhart (2005) have coined a new term for this cellular obstinacy about deciding which stimulus to obey: they call it "weak linkage". This term, not in a very easy-to-read but rather in a somewhat hidden way, gives up the classical model of causality. Cells and genes are not related by a simple cause and effect linkage, as are e.g. the motor and throttle in an automobile. Cells "interpret" the DNA in a kind of consensus procedure, but cells do not "obey". Kirschner rediscovers as an empirical precondition for embryonic development exactly what Varela has postulated by his autopoietic definition of the living as process-of-identity: cells and tissues are autonomous to a certain degree.

They behave as a whole which interprets stimuli but which is not, like a machine, dependent on causal perturbations, or orders. There are no such orders. For Kirschner and Gerhart (2005) the "information" contained in the DNA does not represent an all-determining set of instructions which the cell blindly follows. Rather, it is one stimulus amongst many others, among which are, for example, the state of all the surrounding cells, the temperature, the distribution of signal molecules and hormones, light, darkness, presence of symbionts and other microbes, and so forth.

Because of this inter-relatedness in the way in which a cell is "tuned in" to its environment we have to accept that from the very beginning there is a poetic aspect to the living being. Since a cell acts as an interpretative receptor of outside stimuli, it constantly mirrors every influence that reaches it. This is the start of the endless mirroring process of meanings which an organism is constantly thrown into and in which it relentlessly participates. We can say that the behaviour of one cell expresses the situation of the overall system – and more.

As the system is enfolded in its ecosystem, and via this in the whole biosphere and semiosphere, this mirroring is potentially endless. This is a most remarkable finding. Only if we conceive of the cell as an individual, and hence as something emphatically closed to the environment and not open to its deterministic influences, is the cell liable to mirror anything within its potentially endless reach (Weber 2016:244ff; Brooks and Wiley 1988:355). The cell is the whole in its individual transformation. We might keep this – "the whole in individual transformation" – as a first operational definition of the poetic dimension.

The poetic viewpoint is thus not just a perspective from the outside, a new look upon life as meaning-relating. Rather, the poetic view guides any analysis of the biological processes for it characterizes the way the biological reality is organized on the "inside", so that an individual perspective is born. The poetic dimension therefore provides the central guidelines for understanding biological interactions. Symbols and meaning present the basic framework to analyze all biological processes. Expression is the most primitive reality of life.

Poetics starts with body, not with text. If we describe the life-process as subjective, symbol-generating and meaning-dependent then we already have established the crucial interface between organism and cultural or symbolic processes. Our new description places the organism beyond what Feyerabend (2009) termed the most characteristic problem of occidental philosophy: the mind—body duality.

The Immune-System and the Creation of Self

Let us look at the forming of self through a concrete example. By self I understand a living system that is sheltered from an outside through its own achievement and constant activity. We have seen how the (procaryotic) cell as the smallest really existing autopoietic unit produces itself and by that constructive process creates a world beyond its proper limits. But a single cell is only the most basic unit of integration. In metazoans the whole organism acts as self, but also single organs or tissues can be relatively autonomous. So in what sense can we here speak of self, and how is it brought forth?

Varela (1991:80) distinguishes five layers of 'selves', all of the same processual character: "(1) a minimal or cellular unity, (2) a bodily self in its immunological foundations, (3) a cognitive perceptuo-motor self associated to animal behaviour, (4) a socio-linguistic 'I' of subjectivity, and (5) the collective social multi-individual totality. In all these regions we are dealing with levels and processes where an identity comes about—not as substance, but as movement—and whose fabric of organisation *is* the organism."

We have already discussed the autopoietic genesis of (1). Now let us examine (2). For the distinction of self in higher organisms happens in a concrete manner on a molecular level. The most important system by which this discrimination is done is the Immune system. In a wide sense this system consists of lymphocytes and 'antibodies' produced by them, but also of the cell-recognition and adhesion complexes on the surface of all cells. A common origin of the Immune-system and of the nervous system from a general cell adhesion system is discussed by certain authors (Cf. Edelman 1988). Indeed, a close interconnexion between lymphoid and neural tissues is found in vertebrates.

In metazoans cells have to communicate about how to make up the higher unity they are a part of. Most organisms consist of billions of individual cellular bodies and even of further billions of individuals belonging to other species, symbiotic organisms that together build the being we perceive as unity, as Ilana Löwy (1991:43) observes: "The constituents are closely interdependent and on their own are usually not viable. All symbioses, for instance between nitrogen fixing bacteria and beans, between mycorrhiza and certain forest trees, between animals and photogenic bacteria, and between some wood beetles and fungi, form 'harmonious life units', as do animals, such as the ant colony and ecological units such as a forest. A whole scale of complexes exist, which, depending on the purpose of the investigation, are regarded as biological individuals. For some investigations the cell is considered the individual, for others it is the syncytium, for still others a symbiosis or lately even an ecological complex."

Symbiosis is a much deeper biological regularity than long has been thought. Without it, no multicelluar organism would exist: already the higher (eucaryotic) cell arose probably by fusion of several bacterial cells. Every organism is vitally dependent on its symbionts and viceversa, without their mutual interaction both would perish. Symbionts change their host in such a way that it can not exist without them. The clear separation of host and symbiont holds no longer true. A single organism, so distinctly perceivable by the observer, shows in this view more similarity to a whole ecosystem as to a classical 'monolithical' subject (Margulis and Guerrero 1991).

To escape from this dilemma for a long time it has been postulated that the essence of identity of an organism lies in a Weismannian conception of genetic identity of the germ line. But it has become increasingly more clear that this strong conservation of genetically identical cells for reproduction is rarely the case; and somatic cells show already a strong genetic variation including loss of whole chromosomes. So the idealistic view of unequivocally defining an individual by its material or logical identity fails.

As in cellular autopoiesis where self is only possible through a constant activity, also on a multicellular level the crucial aspect is an active coherence of an underlying heterogeneity. The Immune system in vertebrates is the paradigm of such a heterogeneitiy inside one individual. The immune system is so heterogenuous or "other" to itself, that its main activity consists in a process of bringing forth its own coherence. It is an active process of building a self. In the same way, antigens – what is foreign to the organism – are not simpla that which is outside, as outside in every case needs to be determined through coherence of the immune self.

Immune cells do not start their immune activity after a simple encounter with an antigen, but the antigen must be presented as "other" by the immune system itself. Self is a question of sircular self-production. And "foreign" has to be a part of self to be recognized at all. Chernyak and Tauber (1991:117) write: "Immunity (or organism defined through this system) is concerned with *self*-establishment, and it produces not images of the other, but images (modes) of self-establishing activity; this activity is the organism's Self." Without prior definition of self, foreign could actually not be determined—or: without binding to self, foreign could not be.

The immune system brings forth a multi-layered network of self-assurance (Varela and Anspach 1991:76). The Immune response constitutes only a small part of it. Self is nearly exclusively in contact with self, thereby even mirroring in anticipation any possible contact with an 'antigen': Because the compounds of the Immune system have such a high self-affinity, every agglomeration engendered by an antigen already exists in the body by anti-antibodies, that bind to (in the original view 'antigen-specific') antibodies, and anti-anti-antibodies binding to them and so forth:

> "The end result is that we will always run into antibody classes that will at least partially resemble the incoming epitype E. Stated more simply, the antigen will be able to enter the network to the extent that there is already circulating an antibody with a molecular profile sufficiently similar to it, an "internal image". The antigen ceases to be "determinant" and becomes a small perturbation in an ongoing network. This means that, as which any perturbation in a rich network, the effects of an antigen wil be varied and dependent on the entire context of the network, as is now known to be the case." (Varela and Anspach 1991:78)

Even without contact to an antigen the Immune system is highly occupied with itself. It is a self-referential, positive assertion of a coherent unity. What I "other" becomes understandable only in terms of this coherent unity. One molecule is neither self nor foreign—it is asserted a role only within the interaction of the system. So we have to say goodbye to the view that an antigen agent will be attacked by a clearly seperated self. Foreign is only foreign as the self reacts to it, otherwise it is just not important. And foreign is presented only in the form of self – as the antigen must be processed by the antigen-presenting antibodies. The dynamic state of the system decides to which side a molecule belongs. Self is established in a pragmatic manner in the act of self-assurance and thus creates its own other. What is needed for self is basically a strong heterogeneity– an underlying global non-self.

Self-Construction and Sign-Process

For the living system experiences are only possible mediated through itself. Therefore an outside exists only as presented in the form of self. We see that: (1) For the establishment of an autonomous system a strong heterogeneity is at least as important as a unifying code, (2) the system is defining self and world pragmatically by its inner disposition, (3) the system thus reacts to external inputs not causally but due to its inner states, (4) the system brings forth a sphere of relevance according to its experiences, and (5) the meaning which is thereby created has an absolute-existential dimension.

These criteria—which are valid for all organisms—establish an ontological catalogue to identify 'life'. They satisfy Varelas (1991:80) claim that such a definition "must be sufficiently universal to allow us to recognize living systems as a class, without essential reference to the material components."

The complex network of a living being is able to quit the linearity of cause and effect. Its properties as a network—being empirically observable—are binding any linear genetic causation in a loop backwards on itself. Genes thus only can be viewed as one part of a larger regulatory context. A system that is no longer reacting directly to an input, but only according to its inner disposition, moves out of efficient causality and introduces another type, final causality (Weber and Varela 2002), which is equal with the establishment of an experiencing self. The outer world becomes a parameter in the complicating self-regulation, as seen in the immune system. Many classical philosophical problems, namely that of free will, persist only if the organism is viewed in terms of a deterministic machine and not as an autonomous system.

The transition of the energetical-physical level to the significance level happens at the same moment when the organism manifests itself as a 'whole': then the relation of energetical exchange becomes a relation of signification on the existential background of the organism. The bringing forth of the self creates its world and by this the relevance of this world. In the act of self-assurance self and foreign are (pragmatically) defined. Thus—because what has been exteriorated by the system as foreign is still vitally important for the system (as food, shelter, even mate)—the domain of relevance is created. The relevant world is the system's *Umwelt* established by this process. The dialectics involved is fundamental: Only by self-definition the world as the domain of relevance is established, and only in the separation from this world the self confirms itself. By confrontation with the world that it has excluded actively the subject creates itself.

The making of the self creates an ontological triangle: by self-confirmation the nonself, or other, is separated as *Umwelt*. This separation opens the option, but also the obligation to interact with the *Umwelt* for better or worse. The world gains, in the same movement by which self and foreign are separated, existential significance for the emerging self. This triadic relation can be understood, I think, as the archetype for a sign-process in the sense of Peirce. By this view the self-constitution of the subject *always* is also the constitution of a semiosphere.

Peirce' triadic sign has a characteristic we also find in Kant's *Ding an sich*: it "affects our senses". The perceived object, the signification does not "really" exist as such, but only arises by our touching the *Ding an sich*, that, in turn, will never show in its real nature because we are slaves to our conditions of perception. Only the affecting sign exists on a material- energetic level, but is perceived only in that form which our body makes of the perturbation it causes. The only reality is signification: it springs from an existential need and thus becomes the fundamental grid of experience, the only trace of an underlying complicated ontology. In the organism's perspective nothing remains but this signification. It is the existential condition, and while being existential, it hides its ontological origin and thus becomes the proverbial blind spot on the organism's perspective.

The organism creates a perspective wich changes the world from a neutral place to an *Umwelt* that always means something in relation to the organism. Organisms cancel the neutrality of pure physics and create a concern. Only this organic perspective actually has the status "world", only this is real, because the living can only act on an intentional world.

For an organism the world has a meaning in an existential sense, because the interactions with this world determine its fate. The organism has to keep a "coupling" with the underlying energetical structures. And, as the organism is not causally determined, it must incessantly recreate itself through active regulation. So the background gives sense to the organism's behaviour precisely in establishing a continuous threat to it, as Varela (1991:86) observes:

> "The difference between environment and world is the surplus of *signification* which haunts the understanding of living and of cognititon, and which is at the root of how the self becomes one... There is no food significance in sucrose except when a bacterium swims upgradient and its metabolism uses the molecule in a way that allows its identity to continue. This surplus is obviously not indifferent to the regularities and texture (i.e. the "laws") that operate in the environment, that sucrose can create a gradient and traverse a cell membrane, and so on. On the contrary, the system's world is build *on* these regularities, which is what assures that it can maintain its coupling at all times."

In his 1997 paper Varela goes even further and calls this surplus, which results from the threat that the diverging strands which only active interest can bind together unravel "imaginary dimension". The answer to desperate heterogeneity is imagination. The multitude underlying life must be realized poetically in order to cohere.

Since Feeling Is First

Only the presence of a living being gives sense to the things by transforming them to the stage set of its existential drama. And by this sense all things gain their existential role: their presence or absence for the organism decides about prosperity or defeat, stability or chaos—and this precisely because the organism has to "master the things", because they are so detached from its autonomy that they only *mean* something to him and do not *cause* a behaviour directly.

What is encountered is not neutral, but can only be encountered if experienced through its significance on a gradient between good or bad. In the first beginning perception has no names except the existential feeling of help- or harmful. So in a way we could say that *emotion* is the first unfolding of the world, and maybe we find an emotional background as the deepest underlying structure in all concepts of reality. The first fission of the world, the first discontinuity in the homogenous equilibrium of eternity has no form and no structure, it is nothing but the amorphous cry of highest urgence uttered by the organism: good or bad. When organisms are conceived of as autopoietic systems, *meaning* is one of their fundamental dimensions of existence—in a way it is the true marking point that distinguishes the organic realm from matter.

The background, on which organism is bringing forth meaning, is always existential. For the autopoietic system is all the time concerned with its self-regulation, and only by this self-regulation it keeps a distance from the onset of decay. Meaning arises out of this dependent independence: the living system is not linked causally into its surroundings but nevertheless can exist no moment without it. Because the living being has to keep a distance to matter it gets in a conflict with itself for it precisely *is* matter.

The perspective of a threatened and thus affirmation-dependent organism lays a new grid over the world: a ubiquitous scale of value. Everything the living is interacting with gains in the pragmatics of the interaction an own value related to the amount in which it makes possible the continuity of existence. For an organism the world is at no a time neutral place. It stems from the organism's active self-confirmation that from the beginning it has an interest in itself.

From this perspective arises all the colourful ontological universe we know. This is only possible from the perspective of a fragile being that is all the time threatened by its destruction and thus invents always higher levels of integration. The world's stage without living agents would be a completely neutral place. Only after life has come into it, the world is real in prospering and pain, in joy and misery. Only life is interested in its life as a continuity. By this interest there is an absolute value entering the stage of a beforehand somewhat virtual self-separation of matter. This *absolute meaning* then is the only reliable constant in an organism's life.

The only way we can understand the organism in a non-deficitary manner is to think of it as a relation, a relation between compounds and the process creating these compounds, and thus as a relation between a self and an *Umwelt* together with this self. This latter relation is in a strong sense existential, because it belongs to the network of processes by which the organism distinguishes itself from matter; it is literally deciding about prosperity or failure, about life or death. It is this existentiality that branches out into the finest filiations of possible relations to the world. And it is the organism's paradoxical dependence on its surrounding that lends to that surrounding an unchangeable existential meaning. On this foundation a poetics of the living world must be constructed; a biological theory of meaningful natural signs through the organisation of the living that is realizing its existence.

Nature consists of a multilayered network of interacting selves having historically and autopietically evolved; of selves that behave according to existential meanings brought forth by their very mode of existence; thus of subjects that have an interest in their own existence and whose worlds gain shape by the existential values thus encountered. This definition is situated below every anthropocentric perspective and permits thus to declare a positive inventary of what nature is—although this definition is relatively far from a extensional description like "rivers, meadows, forests". It rather is a minimal ontology of the living. In this form we have found nature, and in this way we also, as a living body, are participating in it.

Empirical Subjectivity

Subjectivity must be viewed as a bodily, material process. It is not a miracle fallen from above, nor a divine exclusivity, nor a rational power only humans possess. It cannot exist without bodies. Without the flesh of a living cell or a multitude of them it is unthinkable. Subjectivity is deeply entangled with matter. Final causation, the goal-directedness of life, is an emotional process. It is not abstract at the first place, not about understandung and structuring reality according to a scheme outside of it. Subjectivity is felt concern. All "higher" manifestations of it are tied to this primitive level that matters most. The history of nature therefore is the history of "embodied freedom."

Maybe the most important aspect of subjectivity, which has to do with all of those other qualities, is that it is shared. It is connected to body, and body is something we share with all other living beings. Subjectivity hence is not private, at least not fully. It has an entirely individual aspect, as my body is only mine, and its characteristics and quirks only belong to it. But as all bodies of all biological species are deeply similar and even intimately related through a common natural history, subjectivity is a perspective which is common among all beings. It is what binds us together.

Therefore, subjectivity is an empirical factor. The empirical objectivity that is so familiar to contemporary science can thus be enlarged by an "empirical subjectivity" – a shared condition of feeling and experience among all living beings. Instead of separating the world into the lifeless objectivity of matter and the arbitrary subjectivity of the unrestricted human mind who is free to chose whatever it wishes, as the common scientific approach has done, we need to see that subjectivity is not free and unrestricted, as it is coupled to the rules through which it emerges as a body. It is empirical, as an embodied subject also is an empirical reality. It is intrinsic to lifemaking, as skin or membrane is.

We share subjectivity to truly become subjects. The world that results from this mutual partaking in different empirical subjectivities is our common objective reality. Because this reality is yielded through empirical subjects, however, it is not plain, but poetic. Poetic objectivity is created by the interplay of empirical subjectivities and is about the individual imagination of the whole through infinite particular lifelines.

Chapter 3
Expression

> *"... the process called poetic invention that mingles breath and sense in a way that no one has explained and no one ever will."*
>
> J. M. Coetzee 1999:149

We have seen that the reaction of living systems to material constraints is the unfolding of a dimension of meaning. Living systems *are* the dimension of meaning of material constraints. Its continued existence for a living system is a subjective absoluteness. But this absoluteness is not only a standpoint of concern. It also manifests in the outside in a concrete, physical form. Meaning becomes visible in body, because it comes into being in and through metabolism, because it *is* body.

A subjective impression therefore is always accompanied by an expression. Autopoiesis happens as a material process which gains material signification. Only by this material process the subjective perspective is granted. In the outward expression of autopoietic systems, meaning appears as form, and as such it reveals itself in a sensually seizable manner. This means that the mode of being of organisms has an irreducible side in which this mode of being becomes visible.

The idea that the gesalt of beings is transparent for their essence has was already known in antiquity. It became particularly prominent in the philosophy of Aristotle. For Aristotle the morphological form of an animal was the expression of its *anima* or *psyché*, its "soul". He considered the soul as a principle of motion inseparable from the bodies of plants, animals and humans, organizing matter into the form of a living being. For Aristotle, the quintessence of organism lies in this organizing form, and to it the living aspires according to its own "entelechy".[1]

From our point of view this concept seems surprisingly modern. "Form" in an Aristotelian sense fits very well with the processual self-identity manifesting itself in the autonomy of form against matter. The Aristotelian concept of embodied soul where the essence of a being is hidden within its flesh, however, for many centuries has been less influential than Plato's view of a world of ideas, of which our earthly reality is only a weak reflection. For Plato the essence of an individual lies beyond its corporal existence. We can see the consequences of the platonic

[1]Aristotle, De anima II, 1,412a; Aristotle, Physik (254b). See also G. Böhme (1992:126), Igambardiev (1999).

A. Weber, *Biopoetics*, Biosemiotics 14, DOI 10.1007/978-94-024-0832-4_3

worldview still in the prevailing scientific attitudes, as e.g. in the concept of the gene as "ideal instruction" shaping the organism from the outside as an object. Aristotelian thinking encounters a renaissance not only in temporary philosophy of biology but also in ethics, in particular where essential or holistic readings are explored (Nussbaum 1986). With the idea of autopoiesis, the aristotelian idea of life as activity from within has gained more traction (see also Chap. 6).

Body as Soul

In the last chapter, I tried to clarify the notion of this "soul-like" interior subjectivity in an analysis of the biological organization. In the following we shall try to see how far this interior meaning dimension, as a cristallization of the autopoietic narrative of the living, also is displayed outwardly.

Organisms react to stimuli from their surroundings in maintaining a dynamic equilibrium as subjects vis-à-vis these external influences, and hence experience these stimuli as meaningful from the view of their existential needs. Meaning becomes manifest for an organism in the perspective of concern. This is always a bodily, material process in space because the living system is existing really in space and time and because the process of the living is about maintaining an identity as matter.

The living dynamic is, as we have seen, always embodied, the balance maintained is physical, and a possible rupture is also a physical one (as e.g. shown in Varelas notion of the "destructive perturbation"). An animal e.g. will react to a shock by flight, to a wound with hiding, and also the bodily reaction to this wound is concrete and hence a visible one (in changing colour in the damaged body area, covering the wound with a new tissue, finally in displaying a scar); a tree on a steep slope will visibly force its roots into the soil, its bark will grow around a barbed wire and show its negative imprint and so forth. There is a general exterior aspect to autopoiesis because organic systems are embodied, because they are reacting as bodies in space to their worlds.

By this the subjective inner dimension gains a dimension that has not been foreseen, but apparently follows as an inevitable consequence from the coupling to the world organic systems maintain. Subjectivity is the principle of organic behaviour and therefore becomes visible in its form. Because we defined cognition as a system's inner-directed reaction to a perturbation and thus as the necessary generation of meanings, the system's history is a display of these meanings, an appearing, a "shining through" of these meanings *as* the system's mode of reaction.

The outside thus is the first place this meaning manifests visibly (and also audibly, palpably). Hence the difference between "inside" in our subjective sense and "outside" in the sense of a morphology is only a distinction between modes of manifestation, or, better, expression. "Subjectivity" in this sense is primarily meant as "concerning the perspective of cognition" and this perspective can be manifest in an inside or an outside. Or rather: it will show in an inside *and* an outside because neither is possible without the other.

Cognition is, as the whole self-realization of the living being, visibly embodied: the transparency of an invisible inside in the outside, a level of expression within the bodily reality of coping with the world. Form is thus necessarily related with what an organism "feels" from its inner perspective. The Aristotelian "soul" as the paradigm of the specific "being such" of a phenotype and its levels of development lies open in every moment. The cramped fists of an hungry baby and then again his sleepy uncertain movements say something about its inner perspective without it being entirely necessary to know how it is feeling "really"—for this "really" does simply not exist in a discursive way, its perspective being its private one. A good portion of it lies open in its form.

For these reasons, our consciousness-centered view misses the point. The central nervous system is embedded in this embodied context as a specialized part of it—but not as something entirely different. The whole reacts to a stimulus—and maybe its expression is all the stronger the less important is the role of a nervous system: think of the immense aesthetic dimension of flowers and anthozoans, the weightlessness of floating jellyfish—the drifting of life itself!—all that is inside turned to the exterior, as feeling turned toward the world.

Mind – or consciousness – on the other hand, has to do with the experience of meaning in an autonomous subject. The question how mind could causally determine the mechanical-causal body turns out to be senseless in the view of embodied subjectivity, as body always wholly is self. And already a simple organism does not react causally to its environment, but, as a system according to its inner states, hence according to meaning.

When, e.g., a protist flees a chemical stimulation, from the protist's point of view there is no reason to admit two different processes, one psychical and one physical. We rather have to deal with a uniform biological process where the flight behaviour is the manifestation of the meaning the substance has for the organism. Consciousness is rather *one* form by which meaning appears, namely the inner form, but it is principally equally valid as the appearance of the same as the exterior, expressive forms.

Thus the question "What may he feel?" in the face of animals (or locked-in persons) is wrongly conceived: What an organism feels we can never know, not even from the near perfect conversation between two fully attentive adult interlocuters. We should rather ask: "What does that mean for him as a subject?" With some intuition meaning can be deciphered from a host of signs—think of the speaking expression of a drying plant, or even the slipper animalcule *Paramecium* that cramps togther its unicellular "body" before it dies from the picrinous acid dripped on the slide.

I am aware that this is only a preliminary discussion of the subject. Intuition here in a first definition is the observation of a correlation between my feelings as a sentient being and the symptoms observed in the other. It also does not mean that I deny the reality or the communicability of human consciousness. It simply means that meaning is transparent already in many ways before—and beyond—consciousness. It means that on a level of biopoetic enquiry, body and meaning are one and the same, only seen from different angles.

Embodied Cognition Is Poetic Expression

Lets turn back to the "expressive" dimension in the pure fact of morphological presence of living things. This sensual presence may well be the "surplus of meaning" Varela talks about—and this just in the sense that an observer does not know how to cope with all these impressions overwhelming him from the realm of organisms realizing their living. It is not always true that nature likes to hide, as Heraclitus said. It equally takes a delight in showing itself. This ubiquitous sensual presence is what Gernot Böhme (1992:131) calls the "ecstatic character" of nature. It beckons forth and affects our—and every being's—senses. "It is easily proven with scientific facts, that nature has such an ecstatic dimension... The concrete nature is full of colours, smells, full of signs and contours, there is nothing that does not make display its presence and that does not contribute with its voice to the great concerto", observes Böhme.

Seeing an inside through an outside can be taken as the second operational definition of the poetic dimension. As the first definition, the "whole in individual transformation" does, it binds together two domains and makes what is inaccessible available through its power of transformation of what we can have access to, and what is, ultimately, like ourselves – or is ourselves – into something not yet known.

Cognition thus becomes a poetic phenomenon of first order. In perceiving natural beings we are able to see their cognitive histories. The inner perspective thus is not basically different from the outer and heeds the same logic of existential feeling. If the ecstatic appearance can say something about its underlying—transformed—cognition, then it speaks also about the inner dimension of the beings which appear in this way. The poetic view therefore is the key to subjective meaning and to the world created by it.

Outside myself the world becomes visible first and immediately only in the poetic dimension. It is thus the most direct access to "knowledge" about the world. But this knowledge is not discursive. It is rather the "archetype of that which is concrete" (Jonas 1973:39). It is as concrete as our bodies are, and as opaque. And it is as existentially concerning – and as contagious with its inner concern and aliveness. Sensually experienced expression is not neutral information but meaning.

This brings me to another consequence from the logic of the autopoietic system. Successful Autopoiesis is the achievement of life's self-assertion, it is thus the materially successful gesture of saying "yes!" to itself. Therefore what is expressed in the ecstatic aspect of the living must be continued existence under the aspect of its success. But that means flourishing, flowering: Already the most simple form of life in its visibility shows an ecstatic moment in the most emphatic sense of the word, overflowing, self-transgressing. Life as self-assertion is life as joy.

Thus already pure and simple life is exuberance—and this results just from the autopoietic process of creation. In living, life is celebrating itself because only as an emphatic gesture of self-assertion it can be life, since it always has to confirm its existence emphatically to assure its continuation. The generous shining of flowering trees in May: Can we postulate that the beautiful has its logical place as the necessary formal counterpart of achieved Autopoiesis? That the realization of the living not only has a beautiful component or can be treated under an aesthetic aspect but that it rather is beauty in its essence?

Here I am more or less on the same page as Böhme (1992:131) working on his "aesthetic theory of nature". He proposes to apply "exactly that theory of nature that uses the ecstatic aspect of natural things as methodological guiding principle".[2] Böhme founded his analysis on a discussion of the Aristotelian view of *aisthesis*. For Aristotle the *aistheton* was an intermediate between the experiencing subject and the thing experienced. This is just the case with our poetic "display" of cognition. It must also always be understood as the result from an interaction between organism and the surroundings. What is expressed in form is, as meaning, not only the subject's history and that of its phenomenal world brought forth in an idealistic manner totally detached from physical, hence "real" constraints (see also Chap. 8).

The surroundings as substrate and source of perturbations of the individual's boundary do not have an objectively fixed form before this encounter, but they nevertheless "somehow" exist. Its perturbations can vary in degree and force. When such a surrounding gains its concrete gestalt in form of the "umwelt" of a living being there must be a certain realistic relation in the expressive manifestation of the perturbation of surroundings within the living being's form—a relation that can be not without circularity though, because everything "mirrored" that way is accessible only as a phenomenon.

The difference between night and day is such an influence that leaves its traces visbly in the behaviour of organisms: because these are in their functions mostly adapted to only one of these "temporary niches" they become themselves systems with an inner and independent time and thus mirror the exterior change: to pass through the period they are not adapted for they fall asleep and thus pass into their "physiological night". Sleeping is the pause enforced by darkness imagined through a body.

Expression as Mediation Between Inside and Outside

The common phenomenality is created in the interaction between an organism trying to maintain its closure and its coupling to a background that thus concretizes as "umwelt". It is neither the "umwelt" that forms the subject, nor the latter creates the former in a purely constructivist manner, but both emerge from the act of cognition. Gestalt in this sense is no image of the surrounding but rather a sign of the interaction that points mainly to the existential meaning of the *relation* between subject and surroundings.

In the gestalt is enclosed the character of an encounter: It is real as a meeting point of both sides, a "mediation" of body and world through the autopoietic self-realization. The dove's wing is a kind of cognition—an interaction of the subject

[2]For an independent and different version of an "ecstatic naturalism" see also Corrington 1995.

with the influence and the irreprochable demands of the surroundings and thus an existential answer. The shape of a dove's wing, its movement, the sound it makes, its inborn grace, all these are ways a body imagines the air. To experience the wing's complexity by our own senses welds two mysteries together: that of the "realm of the air" and that of being able to fly in this realm.

In natural beings we can decipher unique "sculptural histories of cognition". In their forms we can recognize elements of meanings that have sedimented in the course of individual life history. There exists not only individual memory, but also an embodied morphological recollection of the living. This is an observation also Böhme makes, although he does not trace it to the biological organisation of a living being: "The physiognomic traits of a thing can show, like a human's face, traces of his behaviour and his utterances. They are in a way the ecstasies frozen at the thing itself. To be able to read the physiognomy means to be able to conclude from the forms and lineaments of a thing its past and its habitual forms of expression" (Böhme 1992:139).

In the changes of form which unfold through cognition *hypmomnesis* of creation appears – a morphological memory, in which past events are still alife. In the annual rings that conserve the reaction of a tree to climatic particularities, the past is really present. Differently to an old wooden house a plant is alive also in those parts which belong to the past. Inside the Plane tree, which Buffon planted in the *Jardin des Plantes* in Paris over 250 years ago the eighteenth century is still alive. A tree lives in all times it has passed at once. The same can be said of my wrinkles, grey hairs etc.

Every organism through this becomes the "unlocated sense organ" which Gregory Bateson saw in the flowering meadow. It danced for him an endless dance: the history of its embodiment of meanings in the interplay between the self-realizing individuals and the whole ecosystem, an ecstatic dimension of feeling visible as an the outside. With this view we approach what Bateson understood as "mind": the unfolding of a dimension of meaning by the history of auto-regulation of a living system. A "pattern that connects" all living beings. This "pattern that connects", what traditionally has been called mind, is the presence of inwardness in an outside.

Unfolding the Interior Dimension

In a poetic biology a living being is neither only matter nor only form – it is an embodied subject which can make choices concerning the relationship between these domains. Life is matter and meaning at the same time – "inside" and "outside" intertwined. Metabolism therefore is a process by which bread indeed is transformed into flesh. But the mysticism involved in that conversion is plainly empirical. Matter flows through the subject, which in spite of this fluctuation at each given moment only consists of this flowing matter. Via this atomic flux any subject is linked to everything else.

By labelling the particles involved in metabolism we could show that the same atoms that build ourselves are first in the grains, then in the food, then in the muscles. Matter streams through the body and for fragments of time assumes its fixed substance, only to be shed so that later it can be rearranged in other individuals. In this view the whole physical matrix flows through every organism in any given moment of time. Every living being is a node that enfolds all matter. In that sense every organism is indeed a "centre of the universe", as the Russian writer Solshenizyn had it.

To be able to be this centre, an organism must be fragile, dependent on the influx of high-entropy matter, and, ultimately, bound to die. Only the organism's vulnerability and its final failure – death – engender the form of intentionality or interior perspective described so far. This perspective at the most basic level is the subjective standpoint that collects and combines the organism's experience, seeks out new encounters, avoids negative influences and attempts to prosper. To maintain itself as a definite form in matter the organism manifests a directed tendency, an active interest in its own continuation.

The difference between "inside" in our subjective sense and "outside" in the sense of a morphology is therefore only a distinction between modes of expression. In this sense we can define subjectivity generally as "concerning the perspective of cognition and of semiotic self". This perspective is manifest in an "inside" and an "outside". It will show up in both inside and outside, because neither is possible without the other.

Cognition, as the complete self-realization of a living organism, is visibly embodied. It is the transparency of an invisible "inside" on the "outside", a level of expression within the bodily reality of coping with the world. Form is thus necessarily related to what an organism experiences from its inner perspective. The living body is the ground zero of every possible reality. Living cells and tissues are the forces where the desire for life tries to overcome the dire immobility of matter. Because of this lived poetics, the secret feeling of life is tied to matter which is organized in a way that is able to experience.

Lived subjectivity becomes transparent in matter. Therefore it impregnates matter with an interest and a perspective and gives beauty to it. The subjectivity of the living form inevitably projects itself on the surface to be seen, smelled, heard, and touched. In this respect every living being is an open book, and more: it is an instance of acute, existential, real presence.

Chapter 4
Meaning

"It is the essence of life that it exists for its own sake, as the intrinsic reaping of value."

Alfred North Whitehead (quoted according to Sandra Lubarsky (2014).)

Meaning is a *conditio sine qua non*, without which organic life cannot be understood adequately: meaning is a correlate of the mode of establishment of a self. The dimension of meaning, therefore, is a necessary complement of autopoiesis: Life unfolds, as the other of its physico-chemical existence, a sphere of meaning for a self emerging exactly *in* this process of meaning-creation.

This self thus possesses on the one hand a *form* providing its self-maintenance vis-à-vis an exterior world (created by the very process of self-realization) and maintaining the coupling to the surroundings, on the other hand it displays an inner or subjective dimension, inwardness, the perspective through which the organism is acting towards the world and which thus provides meaning.

Two paradoxes are significant for this ontology: First, the organism brings forth its world by a steady delimination against nothingness, and second, a living being distinguishing itself by this process is nothing substantially different from its environment but rather a regular flux and steady exchange of matter with the surrounding world. Only by the dynamics of living itself, in its incessant self-realization, its identity is kept closed.

This paradox lies at the root of what we will have to discuss in the present chapter. In particular the relationship between causality (as the law governing matter) and teleology (*prima facie* manifest in the organisms self-interest) will be of interest here. In how far can we speak of teleolgy with regard to organism?

This question has characterized much of modern biology. Teleology—the view of nature pursuing goals—has been shunned from biological thinking for at least 150 years. This included the idea of nature itself pursuing goals—the creation of species and their special features –, but also individual beings acting according to

The thoughts in this chapter have first been published in Weber (2001)), Weber (2002)), Weber (2003) and Weber and Varela (2002). Starting from there, the role of Hans Jonas as a pivotal thinker in an embodied philosophy of mind and life has become widely acknowledged, particularly by Thompson (2007) and Di Paolo (2005).

A. Weber, *Biopoetics*, Biosemiotics 14, DOI 10.1007/978-94-024-0832-4_4

intrinsic purposes. How we position ourselves in regard to teleology very much defines the chracter of the biological world we are talking about.

At the origin of the modern reluctance to address goal-directedness in organisms stands the work of Immanuel Kant. Since Kant the old Aristotelian principle of an organismic teleology has been regarded as illusion (Weber and Varela 2002). Biologists in Kant's wake have agreed to call organic phenomena "teleonomic"—a rather weak compromise to avoid the strong connotations of teleology.

This, Kant had claimed, was not the law governing organic reality. Rather it was a bias added by our intellect: Kant insisted that "the innate reasoning categories of mechanistic causality that humans appropriately bring to their analysis of *nonliving* reality were incapable of doing justice to the activities of the living realm. To make sense of life as a phenomenon, human judgement was forced to postulate ... an additional principle of teleological causality".[1]

Kant drew attention to the role of *intellect* conceiving nature: its possibilities are limited in a certain sense. But as it is notorious for his thinking, Kant was somewhat ambigous about teleology. He insisted that organism can not be understood in purely mechanistic terms—what did not mean for Kant that it would definitely *not* work in those terms: about this a judgement was simply not possible for the human mind.

It is this latter point that still provokes a misunderstanding: Kant did not rule out mechanism, but he neither declared it to be "the real reality" beneath the phenomena (Cornell 1986:408). He only was completely pessimistic about the possibility that organic life could ever be explained in purely mechanistic terms—and hence did not believe in the possibility that once 'a Newton of the Grassblade' might do the job (Plessner 1982:247). Kant thus makes "both the mechanical and teleological principles, with respect to organism, mere maxims of inquiry of comparable but not total explanatory power. We simply do not know what, if anything, is 'behind' life, 'causing' its basic purposive quality in some ultimate sense" (Plessner 1982:247).

Meaning as Apriori

To clarify this point and to draw some further consequences from it, we will discuss the biological philosophy of Hans Jonas. Jonas developed a somewhat "pre-autopoietic" concept of organism already in the early 1950s and in an astonishing way precedes the findings of Varela. Jonas, in speaking of "necessity" and "freedom" as the basic features (and paradoxes) of organic life, offers a reading of the problem of causality and teleology that can contribute much to the crucial question in how far the organism is the creator of a "real teleology"—a hypothesis

[1]Cf. Harrington 1997:5. Cf. Cornell 1986:407: "The very recognition of a thing as a living organism, according to Kant's analysis, entails thinking of the end of a sequence of causes as itself a cause."

implied in Varela's concept of autopoiesis. Contrary to Kant Varela can supply a "hard" explanation of the living. Jonas' ontological theorems strongly support this view (see Weber 2003, Weber and Varela 2002 for an in-depth-discussion).

The discussion of teleology is important. In making "meaning" the basic fact of biology, causality in a strong sense has to be discarded. Meaning can only arise if the organism has the choice how to react to the encounters with the outside world. To base the organism's ontology on "meaning" also makes only sense if the existence we are talking about has an absolute value—i.e. if continuing to exist *really* means something for the organism and is not only an intellectual principle added by the observer.

In his philosophy of organism, Jonas puts teleology right in the middle of the stage—but he introduces it rather as an observation in the phenomenological tradition than as a philosophical dogma. For him in organism "holeness is self-integrating in active realization, form is not result, but cause of the sequential agglomerations of matter, of whose sequence it consists" (1992:21). In his analysis of this basic situation Jonas then comes to a host of perspectives on nature that will confirm our findings so far made and help to develop them further on.

For Jonas it is a distinguishing aspect of the organic world, that (1) it exchanges its matter and acts thereby from a subject pole partially independent of the underlying matter, that (2) as precarious existence it is always guided by concern, the need to prevent perishing, and to do this again it is completely dependent on matter whose characteristics are the reason for its concern, that (3) already the simplest forms of life have thus a subjective perspective as a result of this existential need, and that therefore (4) life as such will always be captured in the antinomies of "freedom and necessity, autonomy and dependence, I and world, relatedness and isolation, creation and mortality" (Jonas 1973:3ff).[2]

This is a holistic program that encompasses both protists and humans. It is Jonas' particular idea to legitimize this approach by the results of evolutionary biology: if there is a continuum between highest life forms and simpler ones, there has to be an equal continuum of "psychophysical wholeness", and thus also a continuity of all antinomies of life. The theory of evolution has once for all destroyed the possibility of any dualistic ontology.

If the claim is right that all life forms are related to one another, then there can be not two different phenomenal domains. But from this does not result a biological monism with an adaptionist viewpoint which e.g. Wilson (1984, 1993) takes. For Jonas subjectivity, so clearly experienced by ourselves, is a principle every scientific account has to cope with, and no mere epiphenomenon. It rather is the most concrete phenomenon we can have access to: "If men is a relative of the animals, then also the animals are relatives of man and in certain degrees they must display this inner dimension of which man, the most progressed of his genus, is so clearly aware"(Jonas 1973:190).

[2]translations of originally German texts by me, A.W.

This asssociation of the *conditio humana* with a *conditio naturalis* is of special interest here. For Jonas this holistic perspective is the *conditio sine qua non* for "every philosophy pretending to be rigorous science", as he states in an allusion to Kants famous statement in his Critique of Pure Reason. Every philosopher is a living being and therefore a psychophysical unity like all other organisms. He thus exemplifies life as "being that is able to question itself and to understand itself as cosmical enigma" (Böhler 1994:58). In trying to solve this mystery the particular character of a material thing that is conscious of itself, the enigma of a body subject to causality, and nevertheless experiencing his actions as free, must not be ignored. For in these—seeming—contradictions lies the essence of what we try to fathom.

For Jonas (1973:39) "the living body is the archetype of that what is concrete, and, in so far it is my body, it is, in its immediacy of interior and exterior perspective in one, the only fully given concretum of experience at all. Its factual concrete completeness teaches us that matter in space which we normally experience only from the outside can have an inner horizon, and that therefore the fact that it is extended matter must not necessarily be the totality of its being". Organisms reveal a deeper layer of inwardness within this world that articulates itself by the tendency of matter to form wholes concerned with themselves through feeling.

This statement contains *in nuce* Jonas' complete philosophy of organism and, aside with it, his whole ontology, assuming that "in the living body the knot of being is tied" (Jonas 1973:41). It is actually by experience of *our* teleology—our wish to exist further on—that teleology becomes a real entity rather than an intellectual principle. It is feeling that says most about life. To give the possibility of feeling priority is a cornerstone in Jonas' philosophy. Indeed, Jonas in his thinking transgresses a theory of biology and ventures into metaphysical speculations of rare force and beauty. He is rediscovering Spinoza's *deus sive natura* and through this is attacking the dominance of the principle of modern science, *deus supra naturam*. Jonas himself has considered his undertaking a strife to purge the modern world from a technocratic form of gnostic metaphysics.

I think that we can follow Jonas' reasoning without taking his metaphysical speculations too much into account (even if they are always tempting). Indeed Jonas takes at the same time the most metaphysical and the most sober and "scientific" position that one can imagine: By locating the basic antinomies of existence within the unsolvable paradoxes of the living body (may that consist of a billion of cells or only one), he says once and for all goodbye to all classical metaphysical "success-stories" (Jonas 1973:4) in the idealistic style which declare death to be an illusion. But he also discards every materialistic utopia of total control over matter (including the secularized varieties of christianity, namely economical technocracy and marxism). Jonas' universal scepticism does not result from a belief in the "end of all great narratives" including that of the subject, but exactly from the opposite: by making the subject—a thoroughly biological, though not classical "scientific" version of the subject—the cornerstone of a naturalized ontology, he proposes that every being is its own grand narrative.

Organic Existentialism and Autopoiesis

Let's remember the almost simple criterion Varela offers for a definition of life: "Organisms are fundamentally a process of constitution of an identity" (Varela 1997:73). It is characteristical that also Jonas offers such a criterion: for him, this is metabolism: "Metabolism can very well be considered as definig quality of life: every living being has it, no nonliving has it" (Jonas 1973:83). Jonas and Varela do not differ much in this point: as a consequence of metabolism we find just that strange and immaterial "constitution of an identity".

Metabolism means that organims are in a steady flux of matter: their substance in no moment remains one and the same, but they constantly maintain their identity. The identity is maintained exactly by the underlying change: "In this strange process of being for an observer the particles of matter that make up the organism in each moment are only temporary and passing contents. Their identity does not converge with the identity of the whole through which they pass. But it is exactly by the passing of alien matter as part of itself that the whole maintains its spacial system, the living form. From a material point of view it is never the same and although it keeps its identity exactly by not keeping the same matter. If it ever will be the same as the sum of its matter it has ceased to live . . . " (Jonas 1973:120).

For Jonas the organism's identity thus is something very different from its material composition. Not during short moment the organism as a whole is materially identical with itself. This observation is first nothing much more than a simple fact of classical physiology: Its reaction cycles are organized in a way that food delivering energy (in form of energy-rich chemical bonding) and body-substance, becomes itself incorporated in the cycles *constituting materially the body substance*, whereas other, "used" compounds of the cycles, and hence parts of the body, are excreted.

What makes me up now *materially* has been crops on the field; it is the same atoms, in living I incorporate the world and give back parts of my own body—an eternal real transsubstantiation. From this follows that an organism has no similarity with a machine driven by fuel whereby all of their parts remain unaltered (Jonas 1973:121).

This speciality of the ontological status of "organic machines" is central for a theory of organism and thus a point of departure for all further reflections. The dichotomy between process and substrate—that is bridged factually in every moment by the total identity between process and substrate—contains and prefigures the whole scale of an organic phenomenology formulated by Jonas: subjectivity, intentionality, and meaning. In its nucleus lies the antinomy of substrate dependance and autonomy. This is for Jonas (1973:55) life's main characteristic: its double reality as necessecity and freedom.

Necessity: An organism is dependent on the continuous supply of its material compounds, their incessant input and their unhindered integration. But this necessity again is governed by a principle of freedom. A living system can become an ontological centre that it is able to organize itself into a form, which cannot be explained by the features of the underlying matter (the pure necessity) alone. Form—structure—is what governs matter, although being completely dependent on it and manifesting itself only as a form of matter. This is the unfolding of a real, empirical ontological dichotomy in the being of the organism.

By this active paradox outside gains an inside: "For the first time within being the difference between substance and form, which is a pure abstraction when applied to the inorganic, becomes a real distinction. This implies a complete inversion of the ontological relationship: Form has become the essential part and substance has become the accidential part" (Jonas 1973:125). Jonas' understanding of "freedom" equals Varela's use of the idea of biological autonomy. It is this government of the form—and still more, of the principle of self-realization, self-maintenance, brief, the self-referential imperativ to "be!", that imposes its dynamic on matter.

If Jonas can prove that "freedom" is a certain ontological achievement displayed by a mode of being, he will win the whole battle. If he can prove that the same principle that governs our deeply human, e.g. mental domains, rests in the organic life *per se*, then the problem of the two cultures will be no longer unsolvable but rather a gradual divergence, a mere question of integration.

For Jonas, matter is not a compound in an already structured process, matter is also the factor structuring it. The achievement of metabolism is metabolism itself: "Our first observation is that organisms are things whose existence is their own achievement. That means that they only exist because of what they are doing. Therefore the statement, that the existence of organisms is their own achievement simply means: their activity as such *is* their being" (Jonas 1992:82).

Jonas calls this dialectical priority of form over matter an "ontological surprise". By this expression he wants the ontological factuality of the organic matters to be taken seriously. Jonas underlines that the self-realization of the living is an ontological reality because it is an empirical reality. He argues against Kant's definitions for the obvious "possibility of material systems to be units of the manifold not because of a synthetic apperception whose object they are, and neither because of the pure association of forces that bind their parts together, but by their own power, because of themselves and for themselves" (Jonas 1973:131). With this Jonas excludes the possibility of a "Newton of the grassblade", although a Heisenberg might still do.

In order to naturalize Kant perhaps the only tool Jonas was missing was an empirical theory of auto-organization and auto-production. When Jonas formulated his ideas, dissipative systems and attractor systems which organize the underlying matter and hence form a prestage to autonomy had not yet been described, and so the tendency Jonas was claiming for matter was philosophically much more daring than today: matter, Jonas argued, had the natural tendency to display the "crazy caprice of the living" (Jonas 1973:124–125).

As every organism is only a precarious achievement, death for Jonas is constitutive for life. Without the incessant danger to fall back to the brute forces of pure necessity, there would be no inner adhesion of the life processes. There is thus always a price to pay for the luxury of metabolism that gifts the living with its liberty towards matter.

Life even can be defined as the imagination of an escape from death: "The living organism is mortal not although it is living but because of that fact, in its most original constitution, for of this revocable kind is the relationship between form and matter. Its reality, paradoxical, and in steady contradiction to mechanical nature, is fundamentally a continued crisis whose management is never sure and always only its continuation as crisis" (Jonas 1973:16). Only as antinomy, as paradox, life is conceivable at all.

This reading is not a brute naturalism that makes the survival functions of body the scale of good or worse. With Jonas we can see that the basic style of organic experience can be integrated in the vision of a *conditio naturalis* whose shape has strong similarity with the antinomies of human existence. The organism's autonomy (or freedom) always is center stage. Only if organism is conceived of as a deterministic system we are trapped in the danger of the naturalistic fallacy, imposing value from the outside on something which is neutral. Living beings, however, exist according to embodied values. Their nature is to live according to values. The ontology proposed here is non-deterministic *because* of biological reasons.

Feeling the World as Feeling Values

It is in the role of the other that decides about good or evil, and hence in its eternally possible negation, that the world manifests itself to the living. Only in the light of the "desire" of the living, the world gains structure and gestalt, and those are only understandable in the light of these existential needs. A world without organisms would be a world without meaning; and it is in life's incessant need, that for Jonas a subjective perspective is established. Subjectivity is "the absolute interest the organism takes in his continued existence" (Jonas 1973:134).

Jonas emphasizes the fragility of life in much stronger way than do the authors of the concept of Autopoiesis. For Jonas, a disciple of Heidegger during interwar times in the small university city of Marburg, death has an absolute importance as it contains the seed of life and meaning. In the negation that brings forth the countermovement of imaginative self-assertion, the continuation of existence becomes a primordial value. Only in its light the world gains meaning.

To be in a world for an organism thus first and foremost means to experience its own being according to existential values. Aliveness generates its own central value, because it is a process that intrinsically strives to continue: "The fundamental point of departure is that life says 'Yes!' to itself. In wishing itself to continue it declares itself as a value... May we thus say that mortality is the narrow door through which *value*—the thing adressed by 'yes' entered the otherwise indifferent universe?" (Jonas 1992:87).

Only a small part of all dynamics in the environment are relevant for an organism. All other possible interactions are not part of the world, they just do not show up in the experiences. Only what influences the living dynamics of the organisms is real. It follows that every contact with the world for the organism has an existential meaning. Contact with the world is thus always value—a value defined in respect to the continuation of the organism's existence—independent of the qualities the world will unfold later according to this background.

Throught this, the problem of *Qualia*, the "howness" of experience can be seen in a new light. Everything which there is, is first not an experience of structure but of qualitiy. Every *what* is always a *how*, every physical quality is imposed on an existential quality and the former can only be understood through the latter. The structural quality is the imagination of the existential quality through the imaginary capacities of the different senses.

Every possible experience in its deep organization is rather not structure but value. For an organism meaning is an apriori reality. In the centre of every experience sits an existential nucleus. That means that the world's emerging structures can, at least partially, be traced back to the existential constraints, the organism's constitution, and that the existential mode of each organism will determine widely what counts as experience, so that every species lives in a somewhat different world.

This reading of Autopoiesis and of Jonas transgresses a biological-constructivist interpretation of Kant's Critique of Pure Reason. We can substitute the Kantian Apriori by the biological conditions of experience—particularly by an analysis of the sense organs and the central nervous system. The Kantian Apriori by that becomes a bodily one and rests no longer in a theoretical conception of reason and subjectivity but is transformed to its concrete and embodied version. But in our hypotheses we further transgress this transformation: it is neither the rationality of bodily organisation nor perception that structures the world. Before structure is lent to things, experiences are basically "good" or "bad". Experience is always rational in respect to this organization. But the way it firsts manifests itself is pre-rational: it is there as an attractor or as a repellent.

For Jonas the primordial structure of value manifests in a subjective inner dimension which already the simplest organisms have. Jonas goes a step further than Varela and calls this perspective frankly "feeling". In a way this application of a philosophically difficult term poses the same problem as did Jonas' use of the word "freedom"—and it provides us with the same benefits: the direct junction of our experience with the remainder of creation. It is typical that Varela avoids such notion, and certainly there are a lot of good arguments to reserve the concept for the human sphere. But note that Jonas does not use the word "emotions", and in applying the term "feeling" rather addresses the intense and unreflected affective impulse arising "spontaneously" upon an encounter, or, better, *being* that encounter.

In this way the organic mode of existence is "feeling" rather than calculation—a nondiscursive rationality relying on the biological constitution. The term "feeling" dates back to Whitehead (although he related it to a quite different, panpsychistic metaphysics that avoids the sharp ontological boundary between living and inorganic that in our case is the birthpace of everything else) and then has been used by Susanne Langer as a cornerstone of her metaphysics of art and of the living body.

Jonas substitutes the Cartesian Cogito by a "Sentio" that embraces also the Cogito as a consequence of primordial feeling. The "I think, therefore I am" is exchanged into a simple "I am feeling": "Feeling is the primary condition for the possibility that something can be worth an effort. This something is only able to exist as a data for feeling and as feeling of just this data. The pure presence of a feeling, whatever maybe its content, is eternally superior to its total absence. Therefore the capability to feel is the mother-value of all values" (Jonas 1992:88).

With its possibility to fail the living marks an absoluteness, a primordial value that functions as cast for all other values. This conception is a radical break with a Kantian discursive view of knowledge. In any case Jonas confronts any discursive view with the lived experience of a sentient being that everyone can prove within himself—an applied radical phenomenology that claims empirical status. It is in the absoluteness of "feeling" where lies the commanding strength of that view: The subjective inner perspective is absolute for the one who feels. World, as it is, is feeling, and not structure or intellect.

A Practice of Value-in-the-Flesh

The approach chosen by Jonas has far reaching consequences. We can accept that life is no neutral, value free process, but rather illuminates reality with the sharp light of felt values, of existential meaning, and thus of an emotive rather than rational foundation. Rationality must obey the rational organization of the body—and this is done according to feeling what is good and bad for the sensitive flesh. The biosphere is built on existential concern. We are part of it, and we can feel it, too. Indeed, the ability to feel our own and other's existential needs is our most fundamental biological power.

What we understand of value or values is perhaps the most crucial area of a renewed scientific dialogue. Modern and contemporary scientific thinking has relied on the notion of "value free" research, gathering a picture of a neutral, disinterested universe, ultimately composed of inert (dead) building blocks. Value, valence and values were relegated to private concern, shunning the areas of "good life", psychological well-being, or good relationships from the scientific (i.e. biological, sociological, and economic) inquiry. The claim of "naturalistic fallacy" makes it difficult to formulate even humanistic values, hence leading to the situation in which seemingly nothing can be done against deteriorations many would agree exist.

Still, the scientific description did not entirely go along without the imputation of value. In order to describe the cosmos of natural sciences, and the bio- and socio-sphere particularly, a pivotal point had been found in "development", "evolution" or "optimisation". Systems evolve and become better that their historical precursors which they thus outcompete.

In this manner, value has colonized the void left by the enlightenment refusal to accept non-neutral informations in our picture of reality. The bio-economical story of optimization-through-competition-and-annihilation has become the silent nucleus of a deeply biased value system. This system, however, is not openly acknowledged—a classical case of double bind. Bluntly put: in the absence of the acknowledgement that the creation of value is co-creative with agency, pure use-value has taken over the deep metaphysical assumptions of our culture. It is therefore fundamental that we re-discover value as a pivotal factor and experience of living beings and that we organisms increasingly as subjects, auto-creating and hence bringing forth existential values *per se*.

Feeling and Vital Import

What is felt has been the concern of Susanne Langer's philosophical research. Langer started as a philosopher of logics in the Whiteheadian style of the time of the "Principia", and later turned to a more holistic viewpoint of the symbol as import of embodied feeling. Langer has been recognized for her work of artistic symbolization. It has not been preceived as intensely that as a philosopher of symbolization she is considering the organic world as a whole.

Langer (1953:25) starts with a threefold distinction of signs: there are "symptoms" —what we would call just signs, that means spontaneous expressions like blushing shown by animals and humans alike. There are "discursive" symbols which are conceptual expressions of ideas. But the most interesting category is the "presentative" symbol. This signifies a concept of felt life, or rather, as Langer would have it, *expresses* felt life. Seen in the light of our discussion of Varela and Jonas, the entire body is symbolical in a presentative sense. The poetic dimension is presentation of feeling.

All art works through presentative symbolics. Art thus reaches farther than the discursive sphere and has its roots in our organic foundations. This idea did not receive much attention in the 1950s when Langer was forwarding it in her book, "Feeling and Form". But it is an idea that has importance for a biopoetic theory, as it provides a link between biological and cultural semiotics. In this transition from art semiotics to biosemiotics Langer is guided by the question "what new empirical knowledge of the morphology of feeling can we derive from its image in works of art, and what light can this knowledge throw on the unfelt processes of life and the emergence of feeling, animal mentality, human experience and mind?" (1967:74)

To start we have to ask: What is feeling for Langer? Feeling is nothing extraneous to body. It is nothing only produced by body or contained inside of it. It is the other side of body. Of all holist-biological theories, her idea is most similar to Plessners (1928) view of the "double aspect" of the living, that is on one side matter, and on the other living form. For Langer, biological processes have to be taken seriously as a material and *only* material reality. Life hence can not be different from matter. Thus, the essence of life has to be a kind of "forgotten side" of the material setting. Langer is overcoming the problem with the term "phase": being felt is a phase of biological processes (1972:272).

Feeling is an emergent property that is not exhausted by the sum of the parts: "Constituents of one kind, brought together in a special combination, may seem to produce a new ingredient which is, however, a phase of their own occurrence"

(1967:21). This view joins full-front with Jonas' and Varela's insight, that the process of life is not a material reality, but an imagination happening through matter. It is clear for Langer that feeling is the inward side of a physical, or objective, process on matter: "Any felt process may be subjective at one time and objective at another" (1967:31). Nonetheless, feeling is thus not anchored so strictly within the logic of a living system as it is in Jonas, who is a real predecessor of Autopoiesis.

Langer does not pay much attention to Jonas works although she occasionally makes some critical remarks. But if she does not see so much the structuring function of feeling as Jonas does, Langer understands the form of its structure, once emerged from the movements of organic acts. Feeling is a sort of "turning-to-the-outside" of processes that are normally only "visible" inside. What shape has it? Langer speaks of "forms of growth and of attenuation, flowing and stowing, conflict and resolution, speed, arrest, terrific excitement, calm, or subtle activation and dreamy lapses" (1953:27).

The living process for Langer is organized in acts: small, circular processes joining to larger tissues of organic events and hence forming the "matrix" of the living as a reflexive system of circular acts (1982:90).[3] The matrix is not a material entity, though, but a processual or formal one. It is like the action potentials on the abstractive level of their pure dynamism, and not measured in Volts. This is also the view of the living as a more or less autonomous centre reacting by its own laws. The living matrix or system becomes an active agent, responding to external stimuli according to its own states.[4]

Consequently, organic causation for Langer is not linear, or mechanical. It is rather parallel to the teleological causality we have in Jonas, or, in Autopoiesis. Langer prefers to view an external trigger as a "motivation" (1967:283) that does not determine causally the organic situation.[5] This is quite close to the above discussed "intrinsic teleology" of an autopoietic description of organism. Such an approach also has a strong tendency to culminate in semiotic terms.[6]

Langer also makes considerations about the genesis of values. She has a view of organic experience that is strongly based on the idea that organisms perceive the meaning of situations according to their physiological needs. Situations have

---

[3]"All external stimuli, that have effects on an organism affect the matrix, i.e. the organism as a whole, and through it motivate reactions [. . .]" (ibid.).

[4]"Every distinguishable change, therefore, arises out of the matrix, and emerges as an act of an agent, for such a vital matrix is an agent." (1967:322)

[5]"The only way an external influence can produce an act is to alter the organic situation that induces acts; and to do this it must strike in a phase of ongoing activity, in which it is immediately lost, replaced by a change of a phase in the activity", she writes (1967:283). "Motivation", by the way, is a term also Buytendijk (1958) uses for the same reasons, and, to spot some more relatives, Langer develops her view by discussing Uexkülls *Umwelt*-concept.

[6]This relation sees also Lachmann (2000:154n9) who refers to Maturana.

"values" that represent the primary, i.e. existential characteristics of the perceived world.[7] This is very similar to how Jonas puts things, and it is also quite near to Merleau-Ponty's analysis of the pre-predicative structure of our perception (Lachmann 1999).

To get here, Langer explicitly deals with Uexküll's Umweltlehre, emphasizing the difference of the ambient worlds of different species. As a basic denominator, organisms are dealing with the existential values of stimuli they encounter—not with the things themselves. Uexküll already had tried to describe this. He observed that for a dog trained to obey an order to sit on a chair different objects—and not only chairs—may have a "sitting tone". The dog searches for objects that allows him to take the according body posture. It relates to an embodied category, not to an object class.

Also for Langer, the value concept forms the semiotic nucleus. Values are not invented, nor discovered, but felt: "More and more, then, behaviour—the acts of an organism as a whole in relation to extraorganic conditions—comes to be guided and developed by feeling" (1967:425). Categories are build by their existential relevance as "value" and, according to the kind of organism, only later modified into qualia or, in the case of humans, mental concepts. Value is a facet that precedes the modal discrimination: value lies before the splitting in colors, tones, smells or touches. All these senses have an existential dimension liying deeper than their qualia. From here starts synaesthesia, the speaking of warm colors or sharp tones.

From an existential point of view, the sharpness of a word, of a musical note, and that of a knife are alike. A baby learning to speak calls the hair-brush, the broom and the dog's fur alike, because he feels them with the same sensation of the skin. Symbols act by this basic categorization: they enact values that have the same "import" as the real ones. This is the nexus between biosemiotics and symbol theory (see subsequent chapter).

For Langer, values as they are experienced by the organic layer in humans are reflected best in art. Cultural symbols are presentations of these basic categories. All great art is therefore an approach to organic feeling. Its significance lies in its "vital import", in its relevance to the dynamism of subjective experience. As organic feeling itself does, music—and all the other arts—show dynamism, swelling, rest, tension, peace, hence, the existential values encountered in organic experience. This may often be much more complicated, for art is a very subtle means of expression. But for Langer expression always restes in contact with the organic foundation: "Art is the creation of forms symbolic of human feeling" (1953:40).

Still, there may always be ambiguity in the symbolization of feeling: "The same feeling may be an ingredient in sorrow and in the joys of love. A work of art expressing such an ambigously associated effect will be called 'cheerful' by one interpreter and 'wistful' or even 'sad' by another. But what it conveys is really

---

[7] "[. . .] the primary characteristics which animals see are values, and all the qualities of form, color, shape, sound, warmth, and even smell, by which we would naturally expect them to recognize things, enter into their perceptual acts only as [. . .] values for action" (1972:55).

just one nameless passage of 'felt life', knowable through its incarnation in the art symbol even if the beholder has never felt it in his own flesh [. . .] Even the artist need not to have experienced in actual life every emotion he can express. It may be through manipulation of his created elements that he discovers new possibilities of feeling, strange moods, perhaps greater concentrations of passion than his own temperament could ever produce, or than his fortunes have yet called forth." (1953:374)

Due to its origin in value, and its intermodal nature, "Feeling is projected in art as a quality", and not as a mental concept. Works of art exhibit the morphology of feeling, not by resting on conventional iconography, but by that what Langer calls "living" or significant form—a form that conveys not a content, but which has an effect by transporting a certain value: "Homer refers to the 'wine-dark-sea', although Greek wine is red, and the Mediterranean is as blue as any other sea water. But the translucent blue in the curve of a wave and the glowing red in a cup of wine have a common quality [. . .] This quality is the projected feeling" (1967:106).

Artistic form is acting as a sign on the same expressive level as organic form. That is what Langer means when she speaks of "living form" (see also Böhme 1992). This is also a high-level interpretation to the problem of artistic mimesis: mimesis in this view does not mean to imitate organism or nature but to generate the same effect or value. As feeling is always directly correlated with the motor activities or perceptions it derives from, hence with the expressive gestures of an organic body, artistic expression cannot be separated from its content.[8] From here we can derive a third operational defintion of the poetic dimension: It is a form able to convey feeling.

[8]"But just because the created appearance is all that has organic structure, a work shows us the *appearance* of life; and the semblance of functional unity is indispensable if the illusory tension pattern is to connote felt tensions, human experience." (1953:373) Consider also: "In creating an emotive symbol, or work of art, the creator does articulate a vital import which he could not imagine apart from its expression, and consequently cannot know before he expresses it." (1953:389)

Chapter 5
Core Self

> *"Important is big, affection is warm, intimacy is closeness, bad stinks, help is support."*
>
> George Lakoff and Mark Johnson (1999)

In a poetic biology the subjective dimension is all-pervasive. Any event has a meaning for the living framework of the body. Any contact towards the world comes about in the sphere of subjective meaning. It is anchored in the expression of a body which displays experiences, i.e. the significations of what has happened to him. Subjectivity becomes visible – be the organism conscious of the experience or not.

These ideas might be useful for addressing important questions in brain and consciousness research which still circle around the "hard problem". Chalmers (1996: 4), who invented the term, has defined the "hard problem" of neuroscience thus: "We can say that a mental state is conscious if it has a qualitative feel... The problem of explaining these phenomenal qualities is just the problem of explaining consciousness. This is the really hard part of the mind-body problem" he writes.

For Chalmers the decisive question is how our own subjectivity is related to the world. If we continue further down Chalmers' line of thinking, another question follows: what relates lived subjectivity with consciousness? How does consciousness arise from subjectivity? Who has it? Only humans? Just higher animals? Or all beings? The poetic approach, however, no longer takes the hard problem to be really hard. In truth it is a problem of acknowledging the "soft animal" of the body (Oliver 1986). If embodied subjectivity is the central drive that holds together a living being, then here – and nowhere else – we must look for the trail leading to consciousness.

As Damasio (2000) has shown, consciousness is not possible without processes of subjective feeling in certain brain regions. From a poetic standpoint, this situation is not only caused by brain organization. Rather, the import of feeling is due to the fact that the phenomenon of the living is the embodiment of subjective identity, which already is feeling. Consciousness somehow reflects this subjective identity. There is no subjectivity that is uncoupled from the body. This perspective seems to be the only way out of the deterministic trap in neuroscience that has recently regained momentum. The deterministic argument is illustrated by Benjamin Libet's famous experiment (Libet 1985). He illustrated that there exists a time-lag between

A. Weber, *Biopoetics*, Biosemiotics 14, DOI 10.1007/978-94-024-0832-4_5

the neuronal impulse to raise an arm and the conscious decision to raise the arm: the decision to act is perceived by the subject after the motor impulse that contracts the muscle takes place.

For many the order of this sequence proves that the body enacts decisions mechanically before our mind is permitted the illusion that it has taken the decision by itself. But not so if body and perceiving subject are the same. Then Libet only showed that an autonomous agent becomes conscious of his free decision after the actual action he has taken. It shows that consciousness floats on an ocean of subjective feels – but it does not demonstrate that the body is a machine. In explaining the Libet experiment with our new poetic approach to consciousness, however, we witness the crucial changes living subjectivity brings to the idea of mind: it closes the dualist gap. Subjectivity is no longer outside, but already "inside" as the inner perspective of a living system. The body is a subject, not the "mind" alone.

If the body takes a decision before "our" consent then this only shows that the symbolic mind is not the almighty governor of the body. Seen in this light, Libet's experiment works as a further proof that autonomy is a real trait of living systems. If the material, embodied identity already has a subjective standpoint, then where is the problem with autonomy? It is this embodied autonomy which is symbolized by personal consciousness as one's own autonomous decision.

Meaning and Subjectivity: The Lingua Franca of Life

If we want to get closer to an understanding of consciousness as part of the overall poetic picture of organism, we have to discard the idea of an immaterial mind, not the notion of individual freedom. This freedom rather reveals itself as the organic core reality. It has its roots in the self-realization of the living, not in any purely mental freedom from matter. The discussion about how to interpret Libet's results therefore shows how deeply engrained a dualistic standpoint still is among researchers.

But in order to understand consciousness we have to look more deeply at the centre of subjective concern which guides an organism's behaviour. Things are the reverse of what was previously thought: it is not that if we understand how consciousness has been miraculously brought forth by a mechanistic body-machine then can we understand "subjective feels". Rather only by comprehending how lived subjectivity manifests as a centre of concern will we be able to explain consciousness.

As we have seen above, value guides the self-maintenance of a cell. Therefore value as the import attached to sustaining and unfolding the proper existence is the pacemaker of life. Jaak Panksepp (1998) calls this intentional standpoint the "core self". In the core self, the status of the body (and hence, the body-subject) is mirrored and symbolized. The core self is the existential, and hence also the poetic, anchor of an individual. Here the meanings of the organic processes are collected, integrated and interpreted. This interpretation is not mathematical, but experiential: it is what we perceive as the "subjective feel" Chalmers (1996) is talking about and what he is putting centre stage to tackle the "hard problem".

Yet the core self is not the central processor of a hardwired body. It is not a physical hub but a subjective standpoint. How can we imagine how this core self comes about? The process of valuing the states of the organic self is not simply located at the material level of neuro-endocrinological chemistry. To make an organism feel its own status there must be "neurosymbolical" processes at work (Panksepp 2001). Panksepp argues that the core self results as the symbolic integration of the inner states of a being.

The "subjective feel", the interior perspective, is just what the biological processes mean for this embodied subject on an existential level. Thus, the core self must be understood as the meaning of bodily processes. It is their persistent and irreducibly subjective dimension. Consciousness is the symbolization of this core symbolics to the second power. Consciousness then is not a representation of the body, as Damasio argues (2000), but rather a symbolization of lived subjectivity through the medium of value, meaning and self.

Only the development of an embodied subjectivity allows for successfully coping with the contingent world. For every agent, reality unfolds into a large range of unforeseen events. It is not possible to program an agent according to this virtually endless wealth of reality. The only way to cope seems to assess what is good and what is a nuisance. This assessment is done by subjectively experiencing its meaning in relation to the continued realization of the living.

The representational approach claims that an organism stores each action and the stimuli concerned in some kind of neuronal memory. This means a huge bulk of information must be held in storage, as well as constraining behaviour to the situational landscape it has been designed for. The perspective of a feeling self, however, allows that there exist few algorithms of inborn behaviour, and those act mainly to maintain the integrity of the organism. Coping with the world here means that the organism acts due to the subjective urge to unfold and to protect its integrity (i.e. light feels good; sweetness feels good; a warm embrace feels good . . .). This can be done in any unknown environment, if it is not too disruptive.

The according behavioural patterns follow from the inner organization of the living being and need not be independently encoded. The right behaviour then emerges all by itself in a self-organizing way. It is a form of exploratory behaviour – in the same way that the capillary systems in a vertebrate's body do not grow according to a genetic plan but according to the need for tissues to be sustained with oxygen. The poetic explanation of coping has many advantages over the cognitivist one. It is universally applicable in every surrounding, it demands minimal storage of predetermined information, and it can be changed by very few mutations.

The core self is as immaterial as the power which brings forth the cell's coherence. This power is no additional force which Newton overlooked and which we now have to insert in our physical calculations as another causal factor. It is not the mechanistic *vis vitalis* of the vitalists, as seen for example in the works of the German researcher Hans Driesch. It is no dualist mind coming from outside and guiding the brain via quantum interactions, for example by collapsing the Schrödinger wavefunction, as has been suggested in different forms by thinkers from Eccles (1994) to Penrose (1994) and Hameroff (Hameroff and Penrose 2014).

The Core Self Does Not Represent

Nonetheless this force of coherence is no illusion. On the contrary: the core self is the other, the "inner" side of metabolism without which it would lose its existence. It is the "right" side of metabolism (cf. Ruyer 1977 for the ontology of "right" (i.e. experienced) and "reverse" (i.e. third person) perspectives). The core self opens the ordering perspective, the standpoint of being concerned. This standpoint is genuinely poetical. It does not represent the body in the way in which the master display of a power station symbolizes how much fuel is burnt, how big the energy output is and whether there are any alerts, but rather it is the translation of body into subjective meaning.

The poetics of subjectivity is the only scale in which what is relevant for a living being can be expressed. It is the fitting way to refer to a life process that unfolds, to a poiesis of auto-construction and auto-enhancement that succeeds. Subjective value is the common language of all cells and all organisms, a language of poietics and of poetry. Only decisions taken in this language can have consequences in the world, being retranslated by the body into muscular tonus, tissue tensions and synaptic coherence. This argument reminds us how an artwork, which is also matter and idea, or meaning, at once transmits its power by an expressive gesture.

According to Panksepp we have to follow this paradigm if we want to explain the "lingua franca" of intra- and inter-organic communication. For Panksepp, neuronal impulses create expressive forms or, as Susanne Langer (1953) would have it, forms with vital import. The aroused states of nervous cells and brain regions do not engender subjective propositions by transforming raw data into a digital code as a computer would do. The core self does not represent, it symbolizes. It does not exhibit in a one-to-one-style but it translates. To achieve this translation it tends to invent exaggerated expressions, creative solutions, poetic enigmas. This biological idea dovetails well with the massive evidence of human cultural expression – as, in respect of human subjectivity at least, the poetic works and psychological insights of ages have shown. The core of feeling subjectivity is poiesis and its appearance is to be understood in poetic terms.

The poetic approach to biology permits us to fill an old idea with new empirical validity: to regard ourselves as the living being in which nature's subjectivity gains its own voice. We share the general organic subjectivity, but we can explicitely communicate it, express it, and make it the foundation of a new chain of poiesis, which is less autopoietic than poetic. The particular quality which gives us this possibility of speech in a large sense, of semiotic freedom in culture, can no longer be regarded as something which separates us from nature.

What gives us the most human traits is not a *differentia specifica*, but rather the paradigm of the living, and so the direct consequence of the *genus communis*. We do not stand separate from creation, but are rather situated at its centre. We speak and understand the "lingua franca" of subjective meanings, too. We are a part of this language. We are, to use Gerard Manley Hopkins' (2009) insightful expression, "world inscape": "outside" and "inside" at the same time.

It is our body that consents to the fact that words can make us shiver and that imagined possibilities can make us run away. It is our flesh that brings forth subjective reality. Therefore not only is our subjectivity intimately entangled with our body, but so too is culture. Culture, the characteristic of our species, with which we most emphatically identify, actually does not stand alone but is interwoven with innumerable ways of organic feeling. Any contact with the surrounding world is an act of bodily imagination. We have observed this already in Suzanne Langers view of "vital import": A pointed word can hurt as acutely as a sharpened knife. Both protrude into the inner circle of existential closure which an organic being intends to maintain. Both threaten its integrity.

For this reason we call a hurtful comment "sharp": from the viewpoint of life words too cut an existential lattice. Our embodied perception transforms both the destructive word and the dangerous knife into that universal currency of experience which is valid in all provinces of the domain of life: the meaning for sustained realization of the living. Our language reflects the ecology of those associations. Its images do not observe the logic of space-time relationships between objects; rather they obey a logic of embodiment. Only because this is the case are poets able to explore our subjective feelings within language. What language means is part of the core self.

Core Self and Symbol

The ideas I have developed so far lead to a radical consequence, in that the mind-body problem seems to be overcome semiotically. The correspondence between mind ("feeling") and body is itself already a symbolic relationship. "Feeling" – or the interior subjective perspective – can be read, because meaning and value become visible in the body as correlates of the interior perspective. This is reminiscent of Ernst Cassirer's description of the relationship between symbolic thinking and culture: "The relation between body and soul represents the archetype and first model for a genuinely symbolical relation.... Here neither an inside nor an outside is at the origin, neither a before nor an afterwards, a causing force or a caused effect" (1982: 117; my translation, A. W.).

If we take nature as an assembly of bodies and the body as a part of nature, we can extend this finding. Then "soul" in Cassirer's sense, or "subjective feel" in Chalmers', stretches out as the whole of animate nature lying before us. Nature thus is the outside of an organic inside, hence it is also our inside turned out. Its expression is our intentionality. From the expressiveness lying before me another inside can be experienced as an outside, and a subjective experience of the other, which is no different from me in its basic situation, becomes aesthetically transparent. In that way subjectivity becomes real: the inside, or feeling, is accessible as the cognitive or behavioural dimension of an outside. Both are fundamentally the same, in the sense that they form the matrix of one biological individuality.

These findings let us see our relationship to external nature in a different light. If we stand in a metabolic and semiotic continuum with the remainder of life, then mind, which is the symbolic expression of this continuing subject-body, does not only encompass the meaning of processes that go on in the individual alone. Rather, it always reflects the meaning of being-a-part-of-the-world, and with this, the meaning of social relationships and the significance of the state of the natural environment, which in a very strong sense is our "superorganismic" niche. In depth, the whole animate world is core self.

Several different lines of thought start from these ideas. The symbolic conception of mind can help us to better understand social phenomena and cultural processes (both micro, in small groups, and probably also macro, in big or even global communities). They can provide a pathway for further examination of certain "dispositions" and fashions beyond the structuralist idea that words and common discourses alone structure experience. If "inside" is symbolic of "outside", and "outside" already consists of an interaction of many different other selves, then we have a semiotic layer emerging which is much stronger than mere discourse – and much more related to biological world-making, too.

Cassirer's *Philosophy of Symbolic Forms* is drawing from biological theories conceiving of organisms as subjects, not as mechanisms. Cassirer has been influenced by Gestalt psychology – Kurt Goldstein was Cassirer's cousin and friend. But Cassirer has also drawn extraordinarily on the semiotic biology of Jakob von Uexküll, particularly during the writing of his Essay on Man (1994). Cassirer designs his anthropology as implicit theory of art, while at the same time drawing from an implicit framework of biologically embodied cognition. We can see the same relation between philosophy of organism and theory of art in Suzanne Langer's thinking. And we ind an equal setting in the works of Merleau-Ponty (1964).

Creating the Core Self: "Symbolic Pregnance"

Let us look at how Cassirer approaches the problem how subject and outside are related symbolically. Cassirer enlarges the Kantian transcendental system, thereby substituting the conditions of possibility of perception and cognition by the cultural phenomenon of symbolic forms. All knowledge is in symbols. For us, this seems a very contemporary approach. But contrarily to deconstructivist philosophers, who insist that we cannot transcend the cultural semiosphere, for Cassirer the question why specific symbols gains specific forms remains open.

Cassirer did not cut the relationship to what lies outside of the sign: in fact, he elaborated a theory which explains sense from an embodied experience of sensation. Cassirer stands here on a biosemiotic ground not so different from Uexküll, although in his own references to the ideas of his colleague, he seems not to be aware of this radical proximity (Weber 2004). For Cassirer, symbolic meaning is linked to biology. It is provided by an existential-embodied ground. Things and events gain an 'affective import' (Cassirer 1982, II: 239) inside of us in the process of 'symbolic pregnance'.

Symbolic pregnance is possible because any sensual experience can only happen as a transformation into a symbol which expresses this experience. Things are experienced through sensory perception. Their symbolical value is their value for the experiencing body. At the beginning of the symbol process there are 'primordial' metaphors or symbols (Weber 2003): gestures that arouse the same feeling as is felt in the original experience which they symbolize. Here experiences *are what they mean.*

Symbolic pregnance allows for the entry of total outside into sign. Any biological perception is always real encounter with physical forces. Even the most virtual symbol must be present in the senses to be grasped: a word must be heard, a text seen. It is this physical force which allows the entry of the completely alien into our sign universe. Any contact takes place as an energetic exchange, because being is embodied in time and space.

So even the most alien property can be important for a living body according to its impact. The most unknown does still have import regarding its destructive or fertile influence. There is a deep layer of openness to the other in its complete otherness. Both self and other share the same physical matter of the world, even though both are forever separated by the way the senses construct the phenomenon of the other.

Beyond the Mind-Body-Problem

From this standpoint, we can understand that Cassirer, following Ludwig Klages, could provide a definition of the mind-body relationship which is more radical than most current approaches. Cassirer conceives of the nexus as a purely symbolical process, where 'mind' or 'soul' (or its cultural embodiment) is the symbol of the body, and body the way for the inner perspective, to appear. This interaction is the archetype of any relationship. "Soul" and body, or culture and human organism, show the paradigmatic inside-outside flip connection. Every relationship makes each pole symbolical of the other.

In Cassirer's model of symbolic pregnance, we can observe that meaning oscillates between the two poles of self and of a physical real other. Meaning is not the other translated into self, or a fictional other construed by self, but rather a dialectics between self and other. There is a back-and forth-movement from representing to the represented (Cassirer 1982, III: 236): The determination of the cultural symbol (or, as we could easily add, of the human percept as such) is, in Cassirer's words 'always and necessarily reciprocal, it is a revelation which is issued from Inside to Outside, as well as from Outside to Inside' (1995:77).

To admit this process of symbol generation we have to step out into the communion with real, unchangeable otherness, and we have to see that this otherness is dialectically present already in the principle of subjectivity itself. Symbolic pregnance is possible because any sensual experience provides its transformation into a symbol which perfectly fits to express the experience. Symbolism is a body business. For Cassirer the symbolic understanding of the world is grounded on this basic emotional background, "gefühlsmäßiger Urgrund", as he puts it (Cassirer 1982, II:118). All experiences are valued insofar as they carry gloomy or serene traits that immediately characterize their symbolic import. Cassirer calls this process an "Urphänomen", in reference to Goethe's holist theory of symbols. World has a physiognomic appearance before any interpretation has taken place. It is coloured by a primary accent of value.

In the way Cassirer describes the working of this primary value he comes very close to the biosemiotic account of how meaning arises from a grid of signification which the cognising organism lays out onto the world. As Cassirer states: "Only those sensory experiences are extracted from the fluid stream of perception which somehow show to be related to the centres of will and of doing, which prove to be helping or hindering the whole of the living, which thus prove to be important and necessary" (Cassirer 1983:106). This quotation shows Uexküll's influence in a much deeper way than Cassirer might have admitted. Cassirer speaks of sensory experiences and of the way they become valued as symbols. Already in his terminology he adopts a biosemiotic way of expression.

To Cassirer's eyes, the quality of what is signified is dependent on whether the embodied interaction is good or bad. For a simple organism, we could even say, any experience is an *Urphänomen*. Any experience melts down to this universal existential coin of lived value and is then paid back in due amount for the construction of the sign – or rather for that flash of insight coupling existential value, or vital import to the gestalt of the original situation.

To understand the process of meaning generation proposed by Cassirer we have to go back to the creation of meaning by organism. Because of the living's incessant need of input to keep up the fragile equilibrium of autopoiesis, stimuli gain an existential cognitive significance. This significance is represented to the organism as perspective of existential concern. External influences hence act as signs that have a meaning for the organism's survival.

This may mark also for man, as he is an embodied being, the crucial point where experiences become metaphors, and where a metaphor still has the gestalt of the experience that forged it. Think e.g. of the symbolic power of darkness, which *really* is a frightening situation for a daylight species like man. Darkness marks a family of metaphors which stand for the uncanny, for the gloomy aspects of the soul. The "Gestaltung", the form into which the experience will be molded in order to become symbol, *is* the efficient form of experience which the human organism is undergoing.

In a manuscript, that has never been published, Cassirer discusses the symbolic worlds of certain animal species. Contrary to his statements in the *Essay on Man*, at least some seem to be fitted with a symbolic system which has physiognomic character, as the mythic world order of archaic people has (Cassirer 1944). This idea contradicts the more elaborate (and "official") position in his *Essay on man* where he accepts a "symbolic faculty" solely for man. But this finding even more stresses how important for Cassirer was the relationship between "physiognomy", i.e. embodied experience, and symbol.

With Cassirer, we can construe a genealogy of meaning generation, starting with early human cultures. Their obsession with the facts of nature seems to indicate a state of "physiognomic character", where embodied experiences still *are* what they *mean*. Cassirer thinks that archaic cultures are closer to the "pregnant" experience of meaning as an Urphänomen, and therefore live closer to the poetic source. Even though there might be some truth to such a statement (archaic cultures are basically nondualistic), the way symbolic pregnancy is explored is always mediated, always imagined according to the basic rules of poetical dialectics which already rule the cognition of simple organisms. As recent works (e.g. Morphy 1995) have shown, the semiotic universe of archaic people is rather complicated and at any rate not simpler than ours.

We might learn from this intricacy a lot more of the necessary interrelation between experience of nature and its use to understand human social existence (Descola 1997; Ingold 2000). Archaic cultures do not rest in an innocent "physiognomic" condition towards nature, but they show the interrelation of embodiment and symbolicity with fewer cultural noise. Archaic cosmological systems are not paradigms for the lowest stage of symbolic forms. They nonetheless prove Cassirers link between embodiment and symbol in showing interesting ways how body (or nature) and symbol are intertwined.

Mimesis: How Meaning Arises from Bodies

To understand symbolic pregnance, body must be comprehended as the condition of possibility of expression. As such it lies at the base of any culture. This can be grasped more clearly if we regard how poetic metaphor recalls or even re-enacts primary symbols: It tries to arouse the same feeling by generating vital import (Langer 1953). Metaphorical truth hence is far from being literal, but it has to obey the laws of organic flow and rest, the laws of need, of embodied desire. Only by being imaginary it can be literal. In regard to the core self, also "facts" are imagined, as they are represented as meanings, not as "things". A "warm smile" sends out energy, as a fire does.

There is one fact central to this re-enactment of vital import: what is expressed in the symbol has real properties. The metaphor is partly unchained from its pregnance and starts its own semiotic fecundity. This core phenomenon may account for the independency of culture in reference to its embodied origin. Metaphor entrains all its cultural background framework, but then in its primary core as vital import rolls back onto the subject with existential force.

Obviously, an "as if" gesture can have the same import as the "real thing". Both are only efficient as they are received in the core self in which everything is existential meaning. Living in a family where nobody smiles can chill as the lack of heating can. The reason for this analogous effectiveness may be that all impressions enter the cognitive network melted down to a universal existential value, a kind of synaesthetic protostate (Weber 2003), where "real" and "virtual" is not differentiated and inside and outside not really clearly marked. Symbolic pregnance relies on the core self.

If we look at poetry, we might get a glimpse of what is the enigmatic *differentia specifica* of the *animal symbolicum* seen under an embodied-subject view. The deciding moment in the symbolic (or poetic) achievement is that it does not only arouse a vital import through synaesthetic enactement of feeling. It produces something entirely new, something never heard of which becomes an opaque part of the world itself.

The poetic gesture jumps over the blind spot of cognition with a flash of unexpected light. It knows more than what it was borrowed for. Its wisdom stems from a double source always merging body and culture: the symbol is filled with the embodied wisdom of primary metaphor, and at the same time it brings with itself the material of whole culture, the unfathomed depth of the semiotic system. The symbol is the other which knows of me what I cannot see, but what brings me to life.

We have to accept the independency of culture *just because* it is biologically based. A biological symbol always entrains the newness of the event, the continuing enactment, the absoluteness of interbeing which already presupposes the other in the building of self. From this standpoint we can come back to Cassirer and to his attempt to formulate the genesis of cultural symbols from an emotive base. For Cassirer the relationship between body and soul – culture – is symbolic: it is even the paradigmatic example for symbolicity. From here stems the permeability of body for cultural metaphor.

Biosemiotic mimesis in human terms means to be born in a world of both material meanings and of a priori nonmaterial significations. But with every breath both become part of enactive mimesis, of the interpretative re-enactment of the world. The resulting topography is the organism's own path: twirling lines where the silhouette of mountains and trees and the shaking and shivering of the moving spot become indistinguishable.

Goethe: Poiesis and the Absolute Other

Cassirer was profoundly influenced by Goethe's conception of the symbol (Naumann and Recki 2002) as he was also deeply inspired by Goethe's philosophy of nature, which he took serious as philosophical *science*, not just as poetry (Schwemmer 1997:199). The most important inspiration for Cassirer's view of the symbolic core self through symbolic pregnancy was Goethe's idea of the *Urphänomen*. This is an experience, projected into the perception of a "phenomenon", that in its physical affection of our senses achieves what it stands for as a symbol.

An Urphänomen shows what it is or does. Important examples of this effect for Goethe were the Maidenhair tree leaves with their massive cleft, showing the union of what is "one and double", hence a basic condition of life, and the magnet, embodying attraction and repulsion which happen without being planned. The Urphänomen is the world becoming transparent by being encapsulated in a poetic gesture. When Goethe is proposing that the eye is sun-like, "sonnenhaft", he does not mean that the eye is designed for sunlight – but that in receiving light it somehow mirrors its action by mimetic shape. It interprets the fact that there is light by its own action of shining.

Goethe made the famous statement that in everything there was already an implicit theory about it, and that we needed to listen to that. This could best be done through poetic resonance. Goethe thought that "The highest achievement would be to understand that everything factual is already theory. The blue of the sky reveals the basic law of chromatics. One should seek nothing behind the phenomena, they themselves are the teaching." (Goethe 1998:II,131) We can now reinterpret this from our standpoint: the phenomena are the solution because they lie in the middle of perceiver and perceived.

The phenomena of felt experience are the emotional gestalts of the core self. They are the crossing in which the perceiver becomes entangled with the other to such a degree that he can grasp it without leaving contact with himself. This move potentially links the perceiver to everything else. Now we may understand Goethe's poetic insight that sun and the eye are poles of existential semiotic relations in the field of the poetic space of the world. Here the experience of light is a transformation of the physical source of photon energy giving shape to things and the perceptual construction of a being's own world. Only because I am seen by the sun – as I feel its warm rays on my skin – I am able to see the world. Only because the sun is feeding the food chain with energy I can be. Only because its light lightens the world I have eyes to see this light.

I am the sun's echo and its transformation. Both poles of this relationship are only possible through one another. They are one another's interpretation. The eye is the sun of the world of things, as its shining fills the space with the light of meaning. Light emanates from the eye, because in the eye it is transformed from a mere photon fluctuation to a shining reality. The sun is the source of meaning as its gaze falls onto the mere things and brings them to life, makes them see and be able to be seen.

Among an animal's many body parts, the eye has retained most of this character of infinite mirroring Goethe speaks of as a fundamental part of nature. The eye demonstrates that to look means always to be looked at. One might think also of the paradigm of music which means to understand the newest and the most alien as the deepest part of self, to unfold the intricacy of being-entangled-with-the-world in one of its real extensions: to feel forth through its flesh in a sensual-sensory manner.

We know of Goethe that he took this endless mirroring as not to be analysed. Instead, the highest gift for a human mind was to get in tune with the play of mirrorings of poiesis. This is what in the conception of his time the artist was most apt for. Maybe it is a wise standpoint. At the current moment, it even seems, that the real flesh of the world needs less philosophy than protection. Cassirer (1944:148) accordingly believed that art – and hence the poetic sphere – is not representation but intensification of reality. It dramatises the inside-outside-paradox of embodied subjectivity because it adds to both, inside and outside. Poetics is one version of poiesis, of which the bodies-in-interpenetration are another, melting into one within the core self. In every work of art – and maybe in most genuinely expressive cultural symbols – we stand in front of the living paradox of the union of same and other, outside and inside.

In its sheer coming into being we can observe the enigma – or, as we might feel, the miracle – of the 'quod'. The "quod" is what the medieval scholar Duns Scotus called "haecceitas", "thusness", the sheer presence of something instead of nothing. Robert MacFarlane (2007:31), quoting the old Chinese concept of "Zi-Ran" identifies it as aliveness – as "'self-ablazeness,' 'self-thusness' or 'wildness'", through which the early chinese artists "sought to articulate the wondrous processes of the world, its continuous coming-into-being". The intensification of relations in becoming through relations in becoming stands at the center of aliveness. Intensified experience of life is the experience that an inside has found an outside and that a pure inside is possible as an outside. This is the moment which creates an artistic response through forming an outside such that it can become carrier of an inside. The miracle that there is a possible outside to something which we experience to be our inside, fills the observer with wonder. We are reconnected with aliveness, because we feel with wonder that we are entangled with the poetic gesture by our own manner of existing as living gestures.

Why wonder? This emphasis again has to do with relations. What simply *is* as pure 'quod' primarily is not under human control. More precisely: it is not under technical control nor is it under cultural dominance. It is not under semiotic control at all. This aspect relates to freedom. The pure 'quod', which has become, and not been made, which is *other* in the first place grants freedom. Freedom is linked to the other as the motor of innovation. The phenomenon exists only at the middle ground between self and other.

It is related to self as well as to other, and this relationship is not a 'mere' symbolical one, as the sign-phenomenon is only possible through relations, as are sign, self and other. It is for this reason that feelings expressed via signs are real in these signs, are more than signs, because the reality of self and that of non-self is only possible as the encroachment of one onto the other. This is to admit that the living sign process is always containing something more than just sign. It is containing power. This power is the energy of poiesis which is creating phenomena in the first place.

In the poetic process we find the ontology of subjectivity enfolded. It lies in its transitory state as self-other, which is precarious, but at the same time vibrates with dynamic force. I think that this is what is passed on to the perceiver through the artistic symbol. It is a symbol about symbolization, and hence, if it is an achievement, about embodied subjectivity. For Merleau-Ponty a successful artistic metaphor invites the perceiver to perform the gesture that has created it. The gesture is there, as new: it is an inside flaming out from an outside, but possible only as an inside of an outside, and thus never to be detached from it, and ever being protected by its abyss of wholeness, as is embodied subjectivity itself.

This is the relatedness that opens the world to novelty. It is the chiasm of the two helical strands, leading to innovation by reinterpreting the arrangement of two perfectly familiar constellations. What is given in the middle of the crossing is that which may be called grace. It is beyond self as it is beyond sign. It shines *through* the sign. It is the open door, not the step out of it, an empty mirror, which contains the world.

Chapter 6
Contradiction

When I pronounce the word Future,
the first syllable already belongs to the past.
When I pronounce the word Silence,
I destroy it.
When I pronounce the word Nothing,
I make something no non-being can hold.

Wislawa Szymborska (1996)

The play of life is taking place through the fusion of different spheres, which seem irreconcilable: Inwardness happening on matter, goal-directedness through inert atoms, autonomy through longing. Life happens at the crossover of two mutually exclusive and contradictory realms – in between what traditionally has been called "mind" and that which was customarily labeled as matter. Through the discoveries of modern and current biology this duality has not become attenuated, but rather more perpetrating. Its perplexion has accompanied modern biology since its beginning and even became quite graspable in the very final sentence of the first edition of Charles Darwin's *Origin of Species* (1859:490) with its invocation of the poetic domain through the physical subjects of evolution, the living bodies of organisms, bringing forth "endless forms most beautiful and most wonderful".

Genetic science has added to these paradoxes the puzzle of showing that what before had been the clean materiality of biochemical processes has a semiotic dimension. To speak of the *code* of the DNA means to insert a finalistic, goal directed element into the middle of causal-mechanic reaction cascades. The code paradigm of genetics brought the great breakthrough of modern biology, but it also made it a constant contradiction. Through this, from a conceptual standpoint the living being is split in the middle. It needs to be described in two entirely different ways which are contradicting one another. Code is computable, abstract, binary, and closed. Soma, however, is, though rule-guided, dependent on chance, situative, improvisational, and open to interchange with the surroundings.

Hoffmeyer (1997) has described this situation as "code duality", pointing to the fact that an organism exists by means of two antagonistic descriptions, not independent from one another, but strangely different and not entirely overlapping: the description which is the actual biochemical identity of the living being in the here and now, and its genetical substance. With this, Hoffmeyer wanted to pay

A. Weber, *Biopoetics*, Biosemiotics 14, DOI 10.1007/978-94-024-0832-4_6

justice to a strange constellation. For a long while, since genetic research really took off, researchers totally within the reductionist mindset started using semiotic (and hence necessarily teleological) language.

Hoffmeyer (2014a, b) describes the relation between the two realities of an organism as "genetic scaffolding". For him, the genes are not to be viewed as reigning over the soma and distributing orders (nor as these orders themselves, laid down in abstract binary execution guidelines), but rather as a memory of the somatic process by which its trajectory becomes repeatable. Hoffmeyer argues: "The genome is not controlling ontogeny, it scaffolds it (just as books do not determine culture, but they certainly scaffold it)" (Hoffmeyer 2014a: 23, Kull 2015b). Hoffmeyer develops a perspective on how particularly genes and soma interact in terms of meaning and mutual transformation, and not through the cascading of orders and determination, as was the mainstream view of the DNA-body relation for some decades before. He develops an account of a cohabitation of two mutually exclusive domains which only by being joined together make an organism possible.

Materialistic Reduction and Idealistic Inflation

We must acknowledge how deeply this duality runs through the realm of living beings. It runs so deep, that we could easily take it for a substantial separation. The reverberations of this dualism are intimately connected to the pervasiveness the dualistic view gained from early modern times on: it just felt so natural. This philosophical weight is echoed in Immanuel Kant's question, which he took to be the foundation of his whole thinking quest, to account for how "external experience", the experience of an outside through an inside, a "transcendental ego", in Kant's terms, can be conceived.

In science, and in its philosophy, researchers have tried to deal with this situation mostly by trying to eliminate one of its sides. This mirrors the history of modern thinking which has been predominantly dualist, the positions offered by various schools reflecting possible standpoints in an either-or-situation. Modern biology with its powerful drive to "biochemically explain" all features of organism has incorporated the semiotic findings of genetics without discussing the awkwardness of doing so.

Eliminating one of the extremes, as we know well, has not been successful, however. All to the contrary – the questions of mind, consciousness, qualia, feelings, organismal form and self-creation have not only not been solved, but are returning with new urgency – and only now the connections between feeling, embodiment, expression, and meaning are becoming tackled.

But what if the way to a better understanding of organism – including, which is important, also the understanding of our experience to be an organism – consisted not in a reduction of the complex duality, but rather in its full embrace? What if the confusing double facedness in an organism's description – and in being one – was not a nuisance needed to be flattened out by a comprehensive empiricism, but organic existence gained its particular quality only through this – the contradiction which lies in the synchronous existence of two antagonistic modes of being in the world?

Logical Conflict as the Source of Life

Kull (2015a) has argued that the enigma of contradictory duality must be taken as the departing point for any further analysis of organisms. To Kull, contradiction itself contains the miracles of aliveness. Contradiction is necessary to unfold the categories which characterize organisms. As we will see in the brief overview I will give in the following paragraphs, this notion is not entirely new, but it is totally unheard of in modern biology.

The idea of contradiction, with its ensuing clash of goals and resulting incompleteness, has first been touched upon by Aristotle's scarcely-known idea of steresis, which is "activity by deficency", and which for the greek philosopher deeply characterizes living beings (Michelini 2012).[1] This idea of contradiction as a defining feature of organism has later been taken up by Hegel whose philosophy it characterizes in a way yet not fully understood. We will come back to these interconnections later in this section.

Kull (2015a) does not flatten out the logical problems arising by the embrace of contradiction, but takes them as necessary base for any further conception of living beings. Organisms are intrinsically inconsistent, contradictory, and incomplete. Otherwise they could not be organisms, i.e. alive. Living beings do not function well, which is the key to their functioning and resilience. Only by being contradictory, life can be, only by being vulnerable it can be resilient, only by being ignorant it can learn, and only by being cut off it can get into connection.

We see here again the basic ontological condition of organisms as described by Jonas: As living form, an organism needs to control matter, which nonetheless in every given moment is fleeting, never to be finally brought under the living being's power as an unambiguous substance that would not constantly change.

For Kull, this is not only an empirical finding, but is valid all the more as it follows from a profound logical constellation. In organism, we find a logical incompatibility between the functioning of abstract code, and self-realizing soma. The connection between both is problematic already on a logical ground, not just through practical flaws. It is characterized by a basic incompatibility. And only this incompatibility allows the organism to manifest as a subject. Contradiction provokes intention. Kull (2015b) states: "Without any additional goal defined, the logical conflict or incompatibility itself is the source of intention."

[1]Below I will follow Francesca Michelini's analysis of the role of steresis, deficiency, lack and desire in living beings.

Intention, the perspective of inwardness and concern a living subject builds up, is logically inevitable. And it cannot be circumvented, as it can neither be calculated away. Incompleteness can only be healed through the phenomenological realm of a concrete, embodied subject. Its solution must be imagined in a real body interacting with other real bodies in the world. This solution – life – is non computable, as Kull states. Computational models "are unable to describe the real meaning-making. Indeed, semiosis (i.e., meaning-making) is not computation" (Kull 2015a).

Meaning-making – the unfolding of poetic space – needs a body. But the logics of this body is made from incompatibility and contradiction. From here we can understand that healthy organic functioning cannot mean to get rid of ambiguities, striving to be fully identical with oneself. It rather has to do with how an organism copes with his deepest reality, which is the absence of coherence. Coherence always must be created as an imaginary act happening on matter.

In poetic space, which is the experience of existential meaning, incompatibility manifests – is felt – as need. Need is the desire to make coherent what is not, the desire to fit together two realms of different consistency, weave them into a fabric that shines for a moment before it curls, wrinkles, and then becomes ripped apart. We, dwellers of poetic space, living subjects, animated matter, experience need – the manifestation of lack and contradiction – as the ground zero of our existence. We can experience incompatibilty from the inside, and we are rather successful in transforming it through temporal improvisations. In this respect, any organism is the only valid – and most simple possible – computation of itself, focusing not on a contradiction-free logical cascade but on truly working with incompatibilities.

Contradiction as Desire

When logical contradiction becomes embodied, it manifests as need. Need is contradiction from the inside. Incompatibility, need, and purpose are born from the same source. They come into being together, as the facets of one and the same thing. This was clearly seen by Hegel who on this finding based a central part of his phenomenological thinking (albeit this original connection became largely forgotten in the last 200 years). The paradox of dialectics – the contradiction and the ensuing strife to heal it – for Hegel stems from the situation of the living body.

Michelini has rediscovered Hegel as a philosopher of organism. Hegel's ideas fill some gaps we encounter when we try to understand the paradoxical nature of embodied life. To Michelini, Hegel's idea of living beings revolves about what he calls the "activity of deficiency" (Michelini 2012): "For Hegel", she argues, "life itself is imbued with contradiction because it is inextricably bound up with what it lacks: its identity is at one with its negation". An organism's identity is its material status at a given time. But this cannot remain constant, as only through change (embodied in metabolism) life can continue.

Contradiction for Hegel is real, and not a mere logical concept: "It is said that contradiction is unthinkable; but the fact is that in the pain of a living being it is even an actual existence" (Hegel 1969:770). As an experience, this contradiction manifests as need. Need is felt purpose, the desire to relieve that contradiction. Hegel states: "Need and drive are the readiest examples of purpose. They are the felt contradiction, as it occurs within the living subject itself; and they lead into the activity of negating this negation (which is what mere subjectivity still is)" (Hegel 1991:280). If we translate Hegel's somewhat heavy prose ("negating the negation") in terms of embodiment, Hegel says that the organism's answer to contradiction is activity. Subjectivity, the self, arising as part of the organic self-construction, is an activity to counter the steady pull of matter which brings decay.

The unity which an organism creates, and which Hegel calls the "unity of need" is not something inert. It is a continuing process – and as such it is the activity of someone, a self taking a perspective on the world. "It is not an 'empty' unity; rather, it is active, an activity which constantly distinguishes itself by two aspects, the subject and the negative of the subject. Although common thought has it that need indicates dependence on something else, in reality, in a paradoxical way, it is a manifestation of independence" (Michelini 2012:136). This independence, the activity of organism, which happens through profound dependence, is what Kull (2015a) calls "intention", and what Jonas names "freedom".

In a move quite reminiscent of Jonas' dialectics of form and freedom, only need, and hence the activity of lack, allow autonomy to enter the picture, as observes Michelini (2012:137): "Although common thought has it that need indicates dependence on something else, in reality, in a paradoxical way, it is a manifestation of independence: in fact water and food would be totally indifferent to the living being and they would not be able to have a 'positive' relation with it if the living being was not, for Hegel, 'the possibility of this relation'." It is also always possible to miss this relation, so that the organism must be always active to search for it.

Contradiction, desire and identity are bound up in a knot that cannot be untied. Together they form the basic reality of embodied existence with its iteration of paradoxes: Contradiction is needed for identity, identity becomes only real through desire, for identity otherness must be actively excluded. The other is central to the establishment of self. The individual union of an organism is the micro version of the whole in which these incompatibilities coexist and enhance one another through mutual transformations.

Contradiction is sufficient to generate purpose as the felt desire to bring these contradictions to another level and hence make them fertile. Purpose therefore – Michelini (2012) states this very clearly – is always experienced. It is felt. It is even the basic prime experience of being alive. The feeling of inner reality accompanying life hence is: "I want myself as a purpose".

Hegel explicitely connected his early deliberations on identity to organism and living form. Subject is for him a living subject – and not an abstract or purely logical entity. There is an embodied logics of the living, which comes first, and which is, in difference to the logics of Newtonian physics, not exclusive, but incusive and contradictory. But only thus can subjectivity – real, embodied, felt and expressive subjectivity – come to be, as well as the infinite whole can only self-realize as whole through self-limitation: "The subject is a term such as this, which is able to contain and support its own contradiction; it is this which constitutes its infinitude" (Hegel 1970:141).

Tender Transformations

For contradictions to matter, there must be a tendency in matter to ascend to more complexity, and to become self-contained, autocatalytic, and self-productive. I have discussed this tendency, which is the precondition for autopoiesis, in beforegoing chapters. We could describe it as reality's tendency to become self-containing and self-producing. Life turns these tendencies into a real subject of experience, and of expression.

We can see the desire to become self in living beings, and we can identify with it, as we are living beings. The basic experience of the whole which we can have access to in these observations (which are, rather, feelings of connection), is the desire to realize oneself as purpose. This desire is what we need to acknowledge in everything which can give rise to contradiction and its transformation into identity.

The idea of purpose as desire has been introduced into philosophical thinking by Hegel. He gave the term the double twist of being related to life, and to art. In the course of the history of philosophy the relation of the experience of desire, as felt purpose, to an idea of the living has become far less explicit. It was influential in English and German romantic philosophy (the "northern renaissance"), and then, through the ruptures of history, in American transcendentalism and ecophilosophy, and generally within the Anglo-Saxon holist discourse.

In her brief paper, Michelini (2012) brings a crucial aspect of romantic thinking back into focus, as she shows that in Hegel's position life was conceived of as inwardness brought forth through a certain logical necessity of processes. Her work shows that life is "activity of deficiency" (which she calls the "bio-philosophy formula"), hence saying goodbye to any idea of "life" as pre-established balance, and embracing the fact that deficiency itself contains the seed of the imaginative surplus through which wholeness is possible until it yields its own experience of deficiency.

Surplus and lack, in and for and through living experience and expression, reveal themselves to be two aspects of the same thing. Deficiency is, so to speak, a constitutive part of living being itself. "It is therefore in negativity and separation that life unfolds in its fullness and unity... Life is inextricably bound up with what it lacks. Hence one can only speak of 'completion' on the basis of deficiency, or vice versa of deficiency only on the basis of completion" (Michelini 2012:137). Pure surplus – the undisturbed whole – loses the life, as well as total lack does. Living experience is the transformation of one into the other, the transformation of the other of self, contained inside self, into self.

Michelini (2012:138) traces the roots of this thought back well beyond Hegel. She first finds it in Aristotle and his rarely noticed philosophical notion of *steresis*, the idea of privation, that for the greek philosopher is closely connected with the idea of potency. Through steresis, the privation of a positive habit, potency does not become satisfied. In more mainstream western thinking, an ability to do something leads to an act which neutralizes this ability – I can build a house, then dwell in it, and do not need another house. The potency is extinguished by its realization. Fulfilling a lack in this case means that it has gone away. This is the western notion of synthesis and reconciliation. But in life there is a yearning which becomes stronger when its desire is nourished.

According to Michelini, Aristotle's idea of steresis is a potency that through being quenched becomes more profound, more complex, which needs to be fulfilled, but through this only becomes bigger, yielding surplus, but also a deeper lack. Intuitively, this sounds true for all gestures of life: learning yields skills and knowledge, but also the desire for more skills and knowledge; biodiversity does not reach an "optimum state" but simply goes on unfolding, and love is always craving for deeper loving.

If we follow the Aristotelian notion of potency as privation, and the later development of this idea in Hegel, that identity comes into being through lack, we arrive at the opposite of the typical western idea of balance through mitigating need. Then, lack is not longer the core problem to be finally solved, but also the core mystery of wholeness. We can easily see how this western conception of potency always needing fullfilment to become neutralized underlies our concept of progress, of economical but also of personal development (realise your full potential), and anyway the general idea of emancipation from lack and need through material posession. All these ideas are about the strife to be seperate from the conditions of privation. But to be free of privation, to still the lack, also means, as we know now, to be free from the conditions which offer identity.

The "Incompleteness of Life"

Terrence Deacon (2012) has centered his book "Incomplete Nature" around this argument. Not a special quality (or vis vitalis) characterizes life, but precisely the opposite, so Deacon's argument goes. Life is not something added to matter, but something that lacks. The surplus of the living in reality is absence. Living beings strive to realize what they are not yet, they construct a wholeness they are risking to loose in any moment. For Deacon (2012:3), the phenomenon of life is characterized by "absential features". Being an organism is a "phenomenon whose existence is determined with respect to an essential absence".

Deacon, though he never does explicitly, or even knowingly, takes up the motiv Hegel had already introduced in his conception of life, and before him, though to a less elaborate degree, Aristotle, and what also Kull is aiming at when he argues that the discerning criterion of life is its "need". Need happens when that which is central is absent. It is the purpose which comes into being through absence. The "whole" organism is, as Deacon also states, a fiction, which nonetheless guides the acts of a living being. The organic self is what is precisely not there but needs to be achieved. It is present as the unitiy of the desire to achieve a whole. It is a goal which is guiding behavior, ordering it, making it coherent, binding together disparate elements – even literally, as intention to be whole is the force which keeps matter in a self-renewing form. But a goal is not materially present. It is there as an idea, an invisible focus, an absence.

Deacon devotes his whole book to this idea and its various ramifications. Here I can only redraw a few coarse lines. Deacon elaborates one of the most complete accounts of a biology based on paradoxes. He opens up the field of inquiry into the role of "lack" in an analysis of living purposes in a view-changing manner. He does so on many different levels, from the underlying molecular processes up to what we experience as human self. In all he demonstrates that absence is generating a horizon of what to strive for. What is not there but desired, thus becomes the horizon of the living self. That, "which is explicitly absent is me" (Deacon 2012:8).

Deacon acknowledges a deep paradox in the ontology of life, although he does not put it as in the center as radically as do Kull, and Hegel. "Incompleteness" (Deacon) does not immediately sound as contradictory as does "incompatibility" (Kull) or even "negation" (Hegel). But in the end, as organisms are always self-sustaining wholes, to characterize them as necessary incomplete, and to understand the essence of their aliveness as that what is missing in them, discovers is the same paradox.

We find it on many levels, some of which we have already discussed: Varela's "meshwork of selfless selves", Jonas' "freedom in necessity", the contradiction between the sheer materiality of an organism's physical existence and the perspective of inwardness it generates, lawful matter and autonomous meaning, code and soma. Accordingly, Deacon (2012:468) observes that Maturana's and Varela's concept of autopoiesis shows that an organism enacts a logical type violation ("this statement is false"). Autopoiesis is the description of an embodied logical paradox.

Deacon (2012:182) quotes W. Ross Ashby, who regards organization (as in the organization of the biological body) as a restriction or constraint, and hence not (only) as enlarging the degrees of freedom, but also as deeply limiting them. Form enables, but also makes dependent. Here again Jonas' position comes to mind: For Jonas, organization is the meeting point of freedom and autonomy within the living being. For Jonas, form, which is realized as the organism's freedom over the pulling forces of mere matter, is at the same time a neessity, and thus a constraint, a negation of freedom.

Deacon himself deliberates on the physical preconditions for such a paradoxical ontology of life. Does this condition come only about with life? Or does life exacerbate it, and with this, expresses a general condition, which is a fundamental dialectics within the whole? Can the whole only become real by being split into parts? In an attempt to trace the sources of the activity of deficiency contained in the life process, and the idea of the universe as an unfolding-in-contradiction, Deacon (2012:77) refers to Whitehead.

For Whitehead, "physical change necessarily follows from an intrinsic incompleteness in every physical occasion". From here, the road opens to a view of reality in which fixed events do not causally determine one another, but internal contradictions yield meaning and signification. This happens on a physical level, on which, however, feeling, and not causal cascades, set the tone, and "thus envisioned prehension" acts "as the simplest physical exemplar of an interpretation" (ibid.).

There is no space here to go more deeply into Whitehead. But to my intuition the idea of tracing incompatibility back to physics, and to come from here to a process-ontology, is very well compatible with a view of life processes as need–through–contradiction. Deacon seems to go with that. Life processes then, could be substantially express a dialectics which already characterises physical reality.

Privation as an Organ of Perception

The contradictions of embodied existence might be viewed as particular instances of one general polarity, which is the schism between the organism's drive to conserve its ongoing self-production, and the tendency of its building blocks, atoms and molecules, to behave "only" according to the laws of physical chemistry, hence to attain the state of lowest free energy, which is how we technically call the breakdown into dead particles. Again, we come upon Jonas' core polarity, the antagonism between freedom (of the creative self) and necessity (of the physico-chemical rules).

Viewed from another perspective, this is the opposition between self and world, between the whole and the individual, between fusion and self-containment. Life is always mediation between those extremes. An organism, in order to live, needs to integrate both parts in an ever new mixture. It is the transformation of the whole into the self, as we have seen in metabolism, which is the world passing through self *as* self, and it is – again as in metabolism – the trasformation of self into world.

Individual existence means to be a whole as a particular part of one larger whole. This is already the most fundamental contradiction. Living beings are wholes, but they are also parts. There is a basic tension in reality between the taste (or, as Peirce would have it, "habit") of the world to diversify and individualize, and the contrary to this individuality, the totality, which is necessary in order to sustain each individual whole. To live means to somehow take part, and to somehow be suffering from this participation. The phenomenon of life is the fragile imagination of living form through these contradictions.

The existential transformation of these paradoxes is activity of deficiency. Activity of deficiency, in making the lack deeper by responding to it, is ecstatic unfolding. It is learning and growth. Growing by means of transformation does not push away the original deficiency, the desire or lack of the living, but extends it and expands on it: it incorporates it into the expression of desire. Cognition is expression. Incompatibility, the basic material of cognition, needs to be re-imagined on a material basis. Contradiction is not only the logical start of meaning, but also its sensual, embodied appearance. Consciousness and feeling is desire-in-contradiction, and desire is to be perceived through the senses and through the meaningful connections of one's own body.

Contradiction must not be excluded if life is to prosper. Incompatibility is needed, as for the individual only it grants access to the other, which is already a part of self. Still, contradiction is gradual. It must be able to be incorporated. Total incompatibility means destruction. The same scenario applies here as with Maturana's and Varela's idea of perturbations, meaningful encounters the organism makes, which need to be a challenge in order to grant continued connection as well as self-creation, but which can only be incorporated (or imagined) up to a certain degree of disruptiveness. Beyond this, a sign becomes a letal interference.

Life happens on this middle ground as the continued metamorphosis of one side into its contrary. Transformation is the way to integrate incompatibilities. It yields not unity, and not logical cleansing, but interpenetration, and expression. Expression of something through what this something is not but with which it is initmately connected, is the pinnacle of poetic experience – as it is the deep core of relatedness.

Chapter 7
Words

> *"So writing is the method of using the word as bait: the word fishing for whatever is not word. When this non-word — between the lines — takes the bait, something has been written."*
>
> Clarice Lispector (2014)

Life is metamorphosis. This has been observed many times. The german biologist and poet Goethe was one of the most influential proponents of such an approach. This is not just a nice attitude and a hollow phrase. The idea of transformation holds a comprehensive ontology of the living. Goethe tried to demonstrate the reverberations of this idea at his time with his mostly artistic means. Today we can corroborate the general notion of metamorphosis by a very specific picture of the living, of life-as-transformative relationships, and of a world gaining depth through the experiences of individuals. Individuals who are perceiving can be imagined as the whole being able to perceive itself. But to do that, the whole needs to be split up and contradict itself. In this view contradiction is the necessary and sufficient prerequisite for caressing.

For Goethe, the prime example of the deep wisdom of contradiction was, as can be expected, language. Poetic speech, in being "only" words, nevertheless can get into touch with what there is. Metaphors give birth to metamorphosis. Symbols are the way to touch the core of reality by being at the same time pointing to it and totally absent from it.

Language is a prime example of a transformative power. It has not been viewed as such throughout most of recent cultural history, where its powers were thought to reside only in language games and discourse detached from anything such as a "real world". The embodied, and therefore metamorphotic force of language has not gained much attention in sign theory and biosemiotics, although it is inherent in the influential sign theories developed both by Uexküll and Peirce. For Uexküll, the experienced sign cancels, and thus transforms, the efficient sign. For Peirce, the sign process leads to the interpretant, which is the world after the transformation by gaining a new meaning.

A symbol is always the outcome of a transformation and the starting point for the next. It is a major transformative force itself. Therefore, the symbol is an organ of

A. Weber, *Biopoetics*, Biosemiotics 14, DOI 10.1007/978-94-024-0832-4_7

poetic perception. It is the meeting place, where the transformation happens. Here we can see again that poetic reasoning is not a nice embellishment of things totally detached from the empirical business of life, but rather deeply anchored in any of its transformations. Language, and human language particularly so, is not outside of "real reality", but a pertinent feature of it. It is the manifestation of its transformative content. Understanding it as "abstract coding" misses its grounding in the process of transformation-through contradiction.

Language is the ability to "touch" other, to relate to things, processes, emotions, individuals, and through this to turn them into something different. All these items through being related by language, change from objects to fields of imaginative energy, promises of new transformations. Language itself is never an abstract concept. Written or spoken it relies on sensual transmission. Words laid on the characters of the world add a second materiality with according sensual characteristics, which can be velvet, coarse, hollow or smooth.

When Talk Is Touch, and Poetry Perception

Language allows to name the world and through this to own it. But at the same time, being "only" language, a thought, a concept, it exactly forces us to miss the world. Language constrains us to creatively transform everything it touches upon in order to compensate for the necessary contradiction. The power of this transformation starts as a sheer bodily affair. Talking, I can transform things and beings into my own voice and through this incorporate them into my body and transfer them into other bodies. I can roll them around in my mouth and make them leap from my tongue.

This enables me to shower the surrounding world with the newly created bodies of things through the sounds emanating from my mouth. Through words I can change things into something which is not only abstract, an idea, but also something physical again, and by this make them touchable, and give them the power to touch. It is this power that radiates from a beautiful line of poetry which is spoken aloud and seizes the whole body and sends a shiver across the skin.

Through this, language becomes an instrument to replicate this world, which only belongs to itself, to share it with others, and thus to generate more life. The strange paradox that reality is not "in" language but still can be transformed by it, is precisely due to the contradiction introduced by creating a second order level of reference, a new, linguistic scaffold. This scaffold introduces a new level of contradiction. But this contradiction is not different from the symbolic mirroring which occurs in organism all the time, and which produces confusion, and therefore meaning.

I am sending out fragments of the world which has become transformed through my imagination and action, and I am receiving bits of the world with traces of myself in the form of voices, sounds, embraces, variations of touch and entanglement as gifts. I continue the circulation of these gifts by accepting the basic potential of transformation within language.

Things become voices, and thus not only are reduced to the form of signs, but also gain a new embodiment, a new thingness echoing the original speechless one, but also adding to it. Language thus duplicates what it relates to, but it does so in another realm. Duplication and transference: What language really does is generating an abyss in the things themselves. It introduces incompatibility, not much different from the "dual scaffolding" of the body through DNA and cellular processes. And as in biological scaffolding, neither are abstract or "outside". The scaffold is part of what it sustains. DNA partakes in metabolism. Words are sense organs for the world. They create closeness through distance.

Language is a scaffold working as a reference in order to fixate meaning. At the same time it reinvents reality through transforming it. In this, it does not play a different role than all other transformative and hence meaningful processes. It does not exist in a special human realm. But language explicitly elaborates on its character of a scaffold by partly fixed rules, the "grammar of the imagination", as the Italian poet Gianni Rodari (1973) had it.

All scaffolding works via the introduction of a level of incompatibility. It re-invents the world not by substituting its elements nor by reducing them to abstract relationships, but by intensifying the interplay of inside and outside. Human language allows for nominating and thus owning the matters of the world, but only through missing exactly these matters. Language re-creates objects and relationships through something else which is exactly not these objects and structures. It is a system of mutual embodied transformation, creating identity through continuously posing a threat to identity. It follows the "abductive logics" of creation through contradiction.

Men Are Grass

Abductive logics enables a poetic vision which works through participation, through "being of the same stuff". But it functions through an included denial of this participation. It reproduces therefore the underlying tension of identity-in-negation which is the realm of the organism. Objectivity in this view has a poetic aspect. Insights that are not real in a material, physical sense may be valid in a poetic interior sense. Gregory Bateson (Bateson and Bateson 2004) describes this when he compares classical ("objective") logic with a logic that is embodied and subjective.

The classical logical argument that Bateson gives for this goes

"1. Men are mortal,
2. Socrates is a man,
hence
3. Socrates is mortal."

This is an argument we all know well. But this is not the only logics we can find in the universe, particularly if we search for it with the radar of our bodies and our embodied emotions. The alternative is the logics of contradiction. The "poetic argument", which Bateson (following Peirce) calls the logical procedure of "abduction" (in contrast to "induction" and "deduction") does not proceed by abstract comparison of single types, but allows the instantanuous application of several of them, the same way as happens in our perception and feeling. To understand abductive logics, we must be a part of the world and we must be able to relate. Bateson gives an example comparing the logical relationships between grass and humans, which goes:

"1. Men are mortal,
2. Grass is mortal,
hence
3. Men are grass."

Insights of this kind can change our behaviour and in this sense are an influential element of our living reality. They are only possible by sharing the world. But participation at the same time is also a separation, and hence, the creation of a deep contradiction. "Men are grass", on one hand, makes a convincing observation about men, particularly about us all belonging to the same living web of beings – together with all other beings, which are thus not different from us.

But it does so, on the other hand, precisely by incorporating a contradiction. Men are not grass. Men are humans, animals, at that matter, and no plants. "Men are grass" is wrong. It contradicts what we know in order to allow a wider gaze which then not only includes or resolves the opposing parts, but through the threat to identity allows an enlargement of identity. This is the principle of poetic transformation.

As does poetry in a most intense way, all language makes something real through what it is not. But through this power, language, and the poetics of being, do not distinguish themselves from the remainder of the real world. They are rather more deeply entangled with it, and become merely different ways of realizing the same. They become paradigms of creation-in-contradiction. They make something real through what it is not. But also the organism does this in its metabolism that produces itself through striving for an identity that still needs to be realized, through the genes which contain the body by not being identical with it, through the importance of the other for the identity of self.

Language is not different from other systems of embodied transformations. It is bringing forth itself through incompatibility, a nervous system for perception through participation, as the autocatalytic liquid crystal which is the cell, as the self organising system of genetic switches during development, as the immune system, as the family system, as the ecological foodweb. In all these systems, complex relationships arise between partners which through these relationships reciprocally specify their changed identies, and contradict their former ones, and therefore become expressive of a dimension of inwardness, or existential meaning.

Language as a Nervous System

Systems which create novelty through the mutual specification of their participants, or the reciprocal transformation of their interior parts and the outside, can all be classified as nervous systems. Nervous systems transform everything they come into contact with through an active transformation of themselves. If we are entangled with reality in a commons of perception, reality as a whole can be classified as a neural system, an organ of perception and expressive transformation.

Through its potential of being a substance of transformation, the linguistic meshwork in itself can be compared to the nervous system of a living being. Language therefore is a sensory system existing outside of our bodies, which we can apply to get into connection with what there is through transforming it. The nervous system of language brings forth itself, but at the same time I am constructing it, it equally contains me and defines what I can perceive.

Seen from that perspective, any living gesture has features of language. And all language is in a deep sense "wild", self-organising, and embodied. But when it becomes explicit, and explicitly used through its existing on a "different" level which hence is incompatible with the purely somatic realm, its metamorphic nature becomes all the more visible. Language is a means of transsubstantiation between bodies and symbolic fragments of core selves, and between different living bodies and different core selves. To make this transsubstantiation possible, lack is necessary. Lack in language is the detachment from matter, its seeming "abstractness", which is in reality a level of symbolic inventiveness.

Its abstractness, its mere existence in "mind" is what makes language incomplete, and hence unites it with the organic realm. For a living organism, lack generates abundance. The ever fleeting matter, the other individual to which the self can never completely connect to, the desire for a future which is precarious, supply the incompleteness necessary to generate subjectivity. Language works through this kind of lack as well. It is incomplete because it does not use material contagion to convey meaning but symbolical suggestiveness, which is entirely characterized through the absence of what it is about, as the poetic example of abductive reasoning has shown.

Language therefore is nothing special to culture, and nothing the human could use as a moniker of his exceptionality. It works with incompletness in order to designate a connection, it creates meaning by imagining a void and a desire through an imaginative surplus. In this, it is no different from body. It does just the same. It is a portion of the living flesh of the creative reality and deeply unites us with it.

The ecology of language thus forms another level of the creative ecology of mind in which we are embedded, another layer of the whole which is self-fissioning, self-regulating, self-perceiving, self-contradicting and continuously creating newness through the transformations required for this perception. Language is wild as is a landscape, and as such it enables us to enlarge our experience of ourselves. For like the nervous system of our bodies, like the butterflies and moths and the blossoms and the autumn foliage transform the stuff they feed on and the beings they nourish, language transforms everything with which it gets into contact through its own mutability and inventiveness. Language touches what we are and through this it changes that which is and that what we are.

Each language has its proper logic, its own resonance frequency, dependent on the genius of the particular bandwidth of expression of a certain idiom, its grammar, its history. Only because language has this proper resonance frequency, its "specific energy", it can answer to a stimulus. But in this answer it always speaks itself. Also in this respect a sign system does not behave differently to a living being, whose sense organs always react to the world with their "specific sense energy" (Johannes Müller 2012). The eye senses every interference, even a hit or a squeeze, with the perception of brightness and of colors. From the encounter between the retina and an obstacle always light emanates. What language touches will become sign, and still remain reality.

A Preliminary Framework for a Poetics of Living Systems

From here we can observe a certain set of basic principles for mutual transformation by incompatibility (adapted from Weber 2014):

1. Perception and communication always means to be touched on a bodily level (light quantums hitting the rhodopsin stacks in the retina, infrared waves making receptors bend and change their physical shape, odour particles locking into molecular grooves and making their topology change).
2. To be touched means to be entangled: A relationship is a process linking self and other in an embodied and physically connected way.
3. Every relationship is a mutual transformation of both of its sides.
4. Transformation is the translation of one side through the medium of the other, like the DNA code is translated through the soma and transformed into body, and as the gestures in a mating ceremony of an animal are translated into the actual mating, breeding and propagation). Relationships yield "transforms" (Bateson and Bateson 2004).
5. The translation activity of living systems can be described as cognition in a nervous system. Examples for these nervous systems are neural networks, brains, immune systems, DNA switches, ecosystems, the planet earth (Gaia), animal behavior codes, the plant rhizome, human languages, musical scores and performances, architecture, drawings, bodily postures and expressions.
6. A nervous system allows to perceive other in the form of the own self. It allows to imagine that, which in principle cannot be experienced as self.
7. Every relationship is a nervous system.
8. A nervous system only works because it is incompatible with itself. On the other hand, it can be understood only through a logics of contradiction.
9. This logics of contradiction is poetic logics. Nervous systems as defined here cannot be understood in an objective sense, but can be connected to and shared in a poetic logic, although they can neither be defined nor reconstructed in it.

10. The experience of self is the expression of one's own identity through the medium of the other. The experience of the other is the expression of non-self in the medium of self.
11. The experience means to focus the world through a sole individual or a sole gesture which is at the same time imaginative and real.
12. This "whole in an arbitrarily small fragment", is the defining moment of poetical experience. Who is seized by it can share the relational and transformational nature of reality.
13. To what degree an experience becomes effective (to what extent a change occurs in a nervous system), is dependent on its imaginative potential. It does not follow from material cause and effect, but from the symbolical meaning for the context of aliveness. Material cause and effect are instances of living relations and transformations, not the "empirical building blocks" of it.

Imagination, Misunderstanding, and Invention

Incompatibility, and contradiction, are the keys. The eternal miss provided by any translation, the satisfaction which is never achieved, but always imagined through desire, are decisive moments of living sensemaking. The uncertainty through which the self can experience itself by the reaction of the other which never can be controlled are necessary conditions for aliveness. The miss is a meeting-in-the-miss. Without this miss the production of meaning is unconceivable, in the same way as without the basic incompatibility between the feeling body and the abstract language of DNA codes neither life would be possible, nor perception.

The whole truth remains unspeakable. Any creation is always imperfect, and it can be creation only thus. Meaning is brought forth in any moment, but not a meaning which is simply valid, easily to grasp (and eventually to rebuild), but the existential meaning of the moment in which a living being self-assembles from a precarious fate according to its needs of that instant.

The imperfection of biological (and perhaps of every possible) creation follows from the incompatibility in the heart of any productive system. Any cell that lives does not only carry its own negation within itself in an intimate way but is totally constructed from its own contradiction, which is dead matter tending to become inert and attain the lowest energy level. But this negation allows that a cell (or plant, or animal) can actually want itself, and that thus a living being can become its own concern. This, becoming one's own concern, produces meaning and the desire for order needed for the next step in the continuity of existence.

But this next step, this next moment of an ongoing continuity, does not resolve the contradictions. It only sustains them for a while, transforming them into experience, into something inward, what is not part of matter, although it totally depends on it. It converts its own negation into the pearl of a moment of pure presence. But nonetheless this presence carries its shadow with it, as the oyster's pearl carries the dull grain of sand in its center, the foreign element which has found its way into the softer inner regions within the shell.

Life is contradiction. Metaphor is contradiction. Poetic and figurative expressions gain their life-giving strength not through nice embellishments but through the power of incompletenes. Jacques Derrida (1990) brought this constellation to the focus in a short essay on poetry (Wolfreys 1998). The essay was intended as a musing about the "poetic degree zero", about that which is the nucleus of poetic speech. In many respects, Derrida's essay is also a statement about aliveness.

What Derrida writes on the poem as the paradigm of poetic-productive world-making could also be said about a living being: It never opens up in totality, it always remains partially unaccessible, it always threatens to disappoint her who tries to enter into relation with it, because the connection to a living being (a creative part of reality on its own accord) is also a separation, an acknowledgement of separateness. Derrida (1990) says: "No poem that is not also just as wounding. You will call poem a silent incantation, the aphonic wound that, of you, from you, I want to learn by heart". This wound is the precariousness of perception which fails, but still is needed in order to connect, and through which sense emerges, before it is known.

To enable this process of connection-in-disjunction anything which is alive needs a willingness to take risks. We can observe this willingness in any wild living being. This observation is precisely one of ther reasons wild nature moves us. With this, it also inspires us: to be more alive, we need to make this willingness to take a risk in being real and in being vulnerable, which the ouzel shows, which every living being can't help but show, an explicit endeavour of our culture (Weber 2013). To be alive, and thus to transform incompatibility in poetic connection, any living being needs openness which again and again confronts fear; openness towards the shadows, the own ones and those which abode in the world. It needs to see with its wound, which is the lack in its middle, which only keeps it going.

Openness means to accept the gaps and fissures without complaining about them. "There is a crack in everything, that's where the light comes in", sings Leonhard Cohen in his song "Anthem". Aliveness needs curiosity towards what can become. It thrives on the acceptance that this world is a terrain of transformation and that there is no change which not also hurts.

Reality Speaks Us

If we – and any living being – can adopt the attitude of being prone to change through being open, and of being in connection through being vulnerable, then to live means to be self and other at the same time, both subject and world, as the living being *is* the process of relation and mutual transformation. Accordingly, two beings interconnected through this relating are not separate things. Each is the relation by itself, but each is so from totally different aspects. Any true relation is already a description of itself. It is outside – a topology of connection – and inside – the expression of what it means to relate, and what changes are necessary through relating.

Through living relations, the world speaks itself. The world becomes its own language for anyone who whishes to listen. This does not mean that the world "in truth" is transparent and graspable in its whole, be it as poetry for a major poetic intuition, or as facts and figures for an empirical mind with serious measuring and calculating devices. It only means that in perceiving we always participate in the world, which thus starts to speak. And through speaking we not only resonate to the world, but we actually transform it.

The poet and ecophilosopher Gary Snyder is aware of this poetic entanglement. He draws a picture of (lyrical) language and its world-producing powers which is very close to the view sketched here. Snyder also fully acknowledges paradox – the need to settle with incompatibility and to even embrace it in order to participate in the circle of living things. The mastery of nonhuman nature lies exactly in this, acceptance of incompatibility, and then bringing the contradiction to fruit in a wholesale way: "Coyote and Ground Squirrel do not break the compact they have with each other that one must play predator and the other play game." (Snyder 1990:4). Being real means acceptance, and then being creative with what cannot be changed, our precarious freedom.

For Snyder (2001), human language, and poetic language at that, allows man the same attitude as wild beings assume from their natural bravery and realism (which stems from the simple fact that they do not have alternatives). Instead of being the epitome of man's difference, and a rational instrument to control the blind chaos of natural impulses, language for Snyder is wild, too. All who have tried to put it into the service of the enlightenment ideal of total emancipation from and perfect control over the necessities, impulses and yearnings of the flesh, have not understood the true core of language, which is that it is not an abstract layer above the real world, but stuff from the world's stuff, and therefore deeply revealing of our participation in its aliveness.

With this, Snyder inaugurates a third position in the current struggle between empiricist and constructivist attitudes. Up to our days in this struggle two canonic positions are predominant – and more so: they are transfixed in a deadlock. These positions are (1) language is a mostly illusionary epiphenomenon of the real underlying material processes obeying the laws of cause and effect, feedback cycles, and functional loops, or (2) language is an arbitrary projection of our fantasies, power structures, and psychological illusions on the world and on the others.

Each standpoint limits the possibilities of knowledge and feeling. It renders them in a way similar to the manner in which the pathologies of somebody who suffers from a deep personality disturbance distort the perception of reality. Both cause an unability to truly feel, the suspicion that true connection is an illusion, the idea that projection is the normal functioning of reality, and that egoism and power struggle are default behaviours. How such a distorted picture of life could become the yardstick of healthy knowledge about reality is an intriguing question. In any case, it does not understand those aspects of deficiency, lack and loss of control, which truly are at the core of poetic worldmaking.

The widespread outlook on the human condition as necessarily distorted overlooks that imperfection is part of the picture, precisely because it invites creation.

Imperfection is inevitable, but it is nothing to resign about. It is no reason to become nihilistic. But the normal outlook on the human condition does not trust insecurity, but confronts it with fear, or a cynical attitude. A biopoetic standpoint, however, does not understand these lacks and contradictions as disillusionments. They are neither faults assumed to be remedied by tech-fixes, nor are they experienced as despair, and countered with an everything-goes-cynicism. They are just the imperfections of the world through which shines the light of endless creation. They are not less painful for that.

It goes to far here to review to what extent the Freudian model of the unconscious with its idea of the subject as a fiction created by repressed drives has been enthroned as the default model of perception, relating, and worldmaking. The mainstream culturalist position deeply relies on the findings of Freud with his idea of the mirror maze we are all apparently caught in and the ensuing scepticism or even nihilism. But it is very well possible, that Freud's objective model of ego-identity did not describe healthy cognition, but a distorted pathology of self. It is also possible that this pathology was (and is) so widely distributed that it seems an accurate description of how human subjects feel: they project instead of connect, they do not know themselves, their feelings are bloated desires, which ultimately must be kept in check by a rigid cultural self-conditioning.

This does not describe the actual picture, but rather a pathology in which natural feeling has been destroyed. This destruction condemns a whole domain of our self-understanding to follow a worldview based not on an accurate descrition but on a distorted account of reality where feeling is doubtful and a problem and where connection is problematic (or not achievable) by default. Doubt of being able to connect still dominates psychology and philosophy ("emotional intimacy, or better, in the struggle therein", see Kyselo 2014). Such a view and its ensuing control of what could count as a useful research topic then must be seen as the domination of a perspective of distorted reality in order to describe reality. This practice has become deeply problematic for our picture of "life". It even has led to the still dominating refusal of any such category as "life", let alone the real experience of it. For a broader discussion see Weber (2013).

For Snyder as a representative of the embodied-poetic worldview, the separation between "real" and "man-made" (or "made up") which both of these positions draw, does nowhere exist. Language is as pervasive as body. Metabolism is already a language (of meaning and concern), and the human language is not detached from the self-organizing and co-creating powers of body, and of the biosphere. In Snyder's (2001) words, "Language reflects the wildness of the universe back."

Human language is a way we set our core self into motion, and hence no different from other, and often unvoluntarily, manners of expressing existential meaning. In this, it is an illustration that life processes happen in a realm which is not yet separated into matter and meaning: as language, matter is tied to meaning, meaning is only possible through a material process of relation, through a direct connection, or even contagion. Both stem from the same source: Speaking, through gestures or words, it is inevitable to speak out being. Its presence cannot be hidden, only be repressed.

The Map or the Territorry

Language *is* world, and it is so only by being transformed, distorted, recreated. For Snyder (1990:18) "language is a mind-body-system, that coevolved with our needs and nerves. Like imagination and the body, language rises unbidden ... we cannot take credits for its power: like clouds, like blossoms". It is a "pattern that connects". We – as all beings – do have unmediated access to being, but this unmediated access does not happen as a substantial possession of objective truth, but through metamorphosis of what is accessed, and of transformation of who is accessing it. By this the way becomes a journey, and the word a song, and the ethics of walking is realised through a dance.

The "third-way-ness" of this idea becomes clear in Snyder's insistence on language going "both ways" (Snyder 2001:1). We are fully in touch with reality, but we can only be so through totally changing everything every time we touch it. Snyder's view is not pessimistic but accepting. He does not take the fact that agency necessarily transforms what it acts upon to be a proof for the inaccessibilty of what is real (and really experienced), but accepts the metamorphotic nature of all connection at what it is: an invitation to be productive and to be in touch. Snyder (1990) sees the self-producing "ordering of impermancence" as the great common denominator of what is real, and finds it in every aspect of the biosphere. Language is part of it. Indeed, it is also part of that what Gregory Bateson (Bateson and Bateson 2004) had called the "pattern that connects" – the transformative dance that connects living beings to a self-organizing universe, and humans to the aliveness which is unfolding all the way.

Snyder's and Bateson's positions overlap considerably, although one of Bateson's core insights seems to counter Snyder's poetic position on language. One of Bateson's major arguments is the difference between "map" and "territory": what an organism perceives – and hence expresses e.g. in words – is not the world as such, so Bateson does not tire to repeat, but only an interpretation of it. Snyder, however, reminds us that the map is in truth part of the territory, and that the territory all the time shines through the cartography we produce of it. It is in there, but through its absence.

The map is not a detached model of the territory, but its poetic transformation. The territory manifests all the time as its own map. The territory is the map, and the map is made of territory, and it is not only map, but also maze. The map, in being the only thing we have to guide our walk, becomes territory. And territory is a maze, from which the creative unfolding can take place. The menu transforms into the meal: "It may be argued that ... the menu is not the meal ... but we need to find ... the way to see with language, to be free with it, and to find it a vehicle of self-trascending insight ... leading back to unmediated direct experience" (Snyder 2001:2). What is real, however, is not objective, measurable, and under control, but profoundly wild. Language, so Snyder (2001:1), is not an instrument of order and control: "The workings of the human mind at its very richest reflect this self-organizing wildness ... Language reflects the wildness of the universe back."

Accordingly, for Snyder, the poetic task is “to see with language”. This seeing is living. To see with language means to acknowledge that all gestures of the living, all bodily moves, all expressions, traces, reactions, actions, in their self-organizing original way, are speaking. Following Snyder, the most profound poetic task therefore is at the same time the most simple act of being alive. When Snyder says that we need to see with language, he means that we must see with the wildness of the world, as language is wild, and the world is wild, and language is the wildness of the world.

Plants and animals do this simply through their being. The language of this being is written down through their bodies. Humans have an additional faculty to do the same thing, language, but understood rightly, this does not seperate them from the wildness of the world, from the self-organizing biological processes, and instead is just another co-creative flavour connecting us to them. Plants and animals, lichens and fungi, we and the others, realize the world through the gestures of our existence. It becomes visible through us, not outspoken, but transparent, in an ineffable way, present-in-absence. Life speaks itself. It is always contagious with aliveness.

To acknowledge the “pattern that connects” means to be entangled in it, to be part in that pattern, and to read what it, other, is like, through what I, self, is like, and viceversa. To be alive means to be weaved into this pattern as part of it, and to be its weaver, at the same time. This is in full contradiction, and still necessary. We can realize this paradox through our embodied identity, which is one form of poiesis. We can speak out this double reality only through poetry, which is another particular form of poiesis. The poetic attitude is our metamorphotic sense, our feeling for transformation, our being the world itself.

Chapter 8
Transformation

"The wildness of a single oak tree is actualized through the wild process of a community."

Gavin Van Horn (2017b)

Near Sant' Andrea, Italy, the sea laps onto the slabs of rock that form the edge of the island of Elba. The waves, smooth as fish bellies, slate gray, white, and aquamarine, shatter into liquid fragments on the rock. In the distance lies Corsica, barely visible in the haze, under a fan of fingers of light. The water that strokes the stones, the boulder rounded and worn away, the wind tousling one's hair, the birds blown by and lost again, come together in a dance. We are commoners of a commons of perception from which our own experiences, our own identities, and those of the world emerge.

Our identities arise through that which we are not: through impressions and touch, through sensory exchanges with that which is stone and water, molecule and light quantum, all of which somehow transform themselves into the energy of the body. All life, from the very beginning, stems from solar energy that is given to all. Our existence in an ecosphere suffused with life is part of a vast commons even before individuality can be perceived.

Each individual belongs to the world and is at the same time its owner, owner of the rough stone speckled by the waves, ruffled by the wind, stroked by rays. All perception is commons, which is to say, the result of a dance of interdependency with the world. The world belongs to us completely, and at the same time, we are fully entrusted to it. It is only through this exchange that we become conscious of it and of ourselves.

The commons of reality is a matrix of relationships through which aliveness is unfolding in ecosystems and history. It conveys the aliveness of biological and human communities from a perspective of metabolic dependency, exchanges of gifts, and the entanglement of actors within their vectors of activity. Living participants bring each other into being by establishing relationships (metabolism, predator/prey relationships, social ties), thus producing not only their environments but their very identities (Weber 2014, 2015).

A fundamental feature of incompatibility and contradiction as prerequisite for life is the fact, that any being can only exist through, on, and with other. But other is not self. Other is the matter flowing through me as my metabolism, the mating

A. Weber, *Biopoetics*, Biosemiotics 14, DOI 10.1007/978-94-024-0832-4_8

partner whom an organism searches, human beings constituting the "society", and finally the world as such to which any individual will fall back on its death-bed.

Being an organism is sharing in continuity. Organisms, the agents which together constitute the biosphere, participate in one another. They form a huge commons, because they are interrelated in such a way that any individual needs others – and in the end, also needs the whole in order to be, and to be an individual part of this whole. Being an organism means sharing, and sharing means to change through the relationships to others.

Interpenetration as Mutual Transformation

Our situation is one of deep transformation through every experience. This transformation simultaneously is a modification of the individual and a modification of the whole. It becomes distinct and visible in an individual only through forms of experience and symbolic expression. Every experience, every interpretation is always an interpenetration which at the same time joins together what is different, and changes both through the other.

Lived subjectivity, as in every living being, is the experience of being a pole of concern, an "inside" that has an outside. It is the experience of participation of the "right side" and the "reverse side" (Ruyer) at the same time; more accurately, the experience of their ligature through this pole of concern which makes up "I". Subjectivity is already intersubjectivity, as it is from the beginning a mediation between inside and outside, where one can only be given as expressed, or symbolised, by the other. Outside, or other, is the symbol of an inside. Inside is the meaning of outside, or the other.

During this process both sides, or poles, meet and then, through this meeting, change. In this, the character of poetic experience becomes mainfest: something emphatically confirms itself precisely by not remaining itself. As in the antigen presentation through a T-helper-cell, other must be presented as modification of self in order to allow a meaningful connection, and self becomes only relevant as modified by other.

The Canadian literary scholar Northrop Frye called this transformative process "interpenetration". Interpenetration is holographic in the sense that something through its transformations by other retains its selfhood and still enters into connection or even becomes totally modified. It becomes the singular-through-the-world. We can, and Frye (1991) did, imagine reality as a chain or an endless set of nested circles of interpenetrations. Connecting to a part of it inevitably means connecting to the whole – all the more as the living body already is connected to this whole as it is "nothing but matter" and hence perfectly dissolves into the whole while still being totally individual.

It is important to point out that transformative interpenetrations – poetic acts or the nuclei of a poetics of perception – happen totally cross-domain. They are not confirmed to either of the poles of classical dualism – mental *or* material, and they are neither attributable to particular segments of reality with which the single scientific disciplines are concerned (thereby happily deepening the cartesian split). Poetic interpenetrations happen in a meaning space which is primary to manifestations such as "embodied", "mental", or "social". They happen in the sphere of aliveness (Weber 2013, 2016; Weber and Kurt 2015), which always has a third person aspect (its structures), and a first-person-aspect (how it feels to be in connection-through-identity). The sphere of aliveness precisely is what is going away if disciplinary boundaries are respected.

This is also partly an answer to a discussion which has taken over cognitive sciences recently, raising concerns about a "body-social-problem" (Kyselo 2014; de Jaegher 2015; Di Paolo and De Jaegher 2016). Embodied cognitive science, while successfully incorporating an embodied first person perspective, now has difficulties integrating the "social sphere". If the identity of self is body, where does the social identity finds its place? The idea of self through transformations, which is the claim of this book, is not chained to disciplinary domains, though. Self is not the material body.

There is no such thing as "the social" in contrast to "the body". Both are manifestations of aliveness which as a mutual enfolding proceeds the generation of precarious identities-in-relation. Aliveness cuts across disciplinary boundaries. The mind-body-problem can even be described as the dilemma of not recognizing aliveness in the world's various dimensions and tissues. The same misunderstanding applies to a supposed "body-social-problem" (Kyselo 2014). Negligence of aliveness always entrails neglect of mutuality, participation, transformation, openness, and freedom.

Miriam Kyselo (2014:1) shares this understanding in her work on the "social-body problem", when she proposes: "Elaborating on Jonas' notion of needful freedom it outlines an enactive proposal to understanding the self as co-generated in interactions and relations with others." Needful freedom, the enactive core situation, is neither in body nor in the social relationship (although Kyselo seems to be slightly biased towards the latter). To understand the living process as the transformation of two poles through a dialectical relationship in which what is needed as continued self is other, and other becomes only perceivable as self, also means to jump over the borders between disciplinary and even categorial domains.

"Needful freedom" therefore not only implies the junction of two opposites of the same logical type (or of the same ontological domain), but it rather means to join these incompatible domains through a relationship of connection through meaning. It is this the "most remote and most intimate relation" conceivable, as Ernst Cassirer thought. "Freedom", if we look at Jonas' wording, may be a physico-chemical description as in "degrees of freedom". But need is something which is experienced from within. Need is even the type specimen which crosses all domains and links them together, fusing an observable behaviour (an individual struggling for something) with an inward perspective on this (I need that) and the according feeling which IS that perspective, and all this with the systematic notion of biological purpose. In need, all this comes together.

This is the reason why Hegel chose need as example of the ontological contradiction, which is always at work in living beings and hence defines them.

Need is feeling, and it is purpose, and it can nonetheless be described in third-person-terms (a "locust's needs for reproduction"). This transgressing of domains, which is precisely the criterion of living beings and their way to open up poetic space, makes it impossible to decide which domain is more fundamental. To believe, as Kyselo (2014:8) seems to do, that the "construction of human identity occurs not in terms of organismic, but rather *social needful freedom*" to my eye leads back into the Cartesian impasse. On a core self level there are no different domains. This is what core self is about: It is the essence of pure existential concern.

Enaction means not only the entanglement of one and another on one level of description. It rather means "deep entanglement" at every level: something is material and symbolical at the same time. Humans have social selves because they have embodied selves; the body is a necessary precondition for the social self to unfold, and the socially derived identity makes the body flourish. The body is the inside of the social identity, its interpretant, its expression, and at the same time the expression of circumstances alien to this.

Identity as Transformation

We have to start by acknowledging that there are existential interpenetrations yielding precarious identities-as-transformations, or individuals-as-wholes, and hence manifesting and desiring poetic worldmaking. This is the basic level. The quality how it is perceived is dependent on the relative perspective, in various forms, levels, and modes of description (third person, first person, structural, emotional etc.). Basic though does not mean "underlying" or "reducible to", so we could equally call it the highest level, the level in which the whole is fractured and awakened throught the part which will be annihilated by the whole.

The diverse levels of identity which are so easily misunderstood as fixed domains of reality (the biological, the social, the cultural) are totally fluid and interpenetrate one another. But they are not random, "everything goes", a colorful multitude, or a "dappled world". They all heed the principle that identities are yearned for and can come to be only through being risked and precarious, thus by interpenetrating one another. Different levels of these classical domains yield more diverse ways of entering into existential-symbolical transformations.

As the outcome of these interpenetrations is only partly foreseable, we can say it is "wild". It is wild in the sense of how Gary Snyder uses the term, in that it follows certain principles constantly, but through following them, it always yields new outcomes which will then be fed into the process again. The interplay of these interpenetrations is self-organizing and influenced by subjects, it is lawful and chaotic, it is majestically objective and deeply individually felt. It is, as Snyder (1990) called it, nothing more, and nothing less, than the "ordering of impermanence".

My position is a monism. But in it there is no substance to which to reduce, only a continuous meeting and metamorphing, a yearning and an answering, a ricocheting of meaning and existential concern which at the same time always comes in the flavour of inwardness *and* in the flavour of matter as the prototypal transformation and prime reciprocal specification. It is a monism of existential concern and according poetic expression, and not a monism of matter or a monism of mind – or sign. It is a perspective of the one through the other. Both spheres are totally real, and both cannot be without one another, making reduction impossible. Reduction of the inside to an outside is as impossible as trying to repress feelings in order to make them go away: feelings are reality, so they cannot be pushed away but only be distorted. Outsides having insides is reality.

To be an outside with an inside means to feel. Feelings are inevitable for an embodied being. They are what connects it to reality. They express what is real in other through the modifications of self, and what is real in self through the modifications of other. Aliveness opens the poetic space in which the cultural, the material, the pysiological, the social domain are not distinguished but are all existential transformations of an individual-identity-through-other and thus through the whole, yielding a felt gradient in the fundamental tension between total connection and complete separation. The degree of existential connection-through-separation, or autonomy-through sharing is what is felt as inwardness, as meaning, and which thus is cognized.

All we do as an act to and through matter also happens in the realm of meaning. This world is entirely outside, but, through its yearning to come more alive, also always inside. This cannot be repressed to make it dissapear, it will only get distorted and this distortion will manifest as a real impact.

Reciprocal Specification

Experience is the meeting place between an individual's striving for self-continuity and other, which in the case of living beings also strives to unfold and prosper. As has been discussed above, other is part of this ongoing self-realization, and at the same time it is separated from self, and even repelled. There is a deep entanglement between individual and world, which makes both inseparable, but which at the same time only allows for the acute experiential pole which is self, and which is pulling back from the world from a feeling of concern-for-self.

This view takes a middle path between two still somewhat prevailing epistemological views: Worldmaking as a projection of a subject on something which remains to be enigmatic. This is the constructivist, and also to a large extent the (French) semiological position, which has incorporated a biased view of the psychological finding that our unconscious makes us feel and act without our knowledge. Culture is the collective enaction of such fantasies. The alternative epistemological view is the materialist or empiricist one: There are not even projections, as any inwardness is only an illusion for making evolutionary agents work better and achieve a higher propagation rate (aka return on investment). Inwardness is not needed. The world is what can be measured and calculated.

When Varela and Maturana started their work in the 1970s, their revolutionary idea of making the ongoing bodily organization the starting point of a biological subject which defines its world, was the first biological theory that did not fall into the empiricist but into the constructivist category. This opened biology for the suspicion of projection, as before had only been the case for the perspective the humanities had on the world.

But how could a solipsistic biological subject, enclosed in its sense functions be in connection ("coupling") with the world? How can it get in touch with other biological subjects? With autopoiesis, biology finally seemed to have left its causal-mechanic standpoint, but only to be burdenend with the problems the culturalistic approaches brought with it.

Varela changed this in his 1991 book "The Embodied Mind" together with philosopher Evan Thompson and cognition researcher Eleanor Rosch. They proposed a position called "Enactivism", which first only slowly gained track but which in the last years has matured to a mainstream approach in cognitive science, relegating representational and computational views somewhat to the past. Varela was fond of using the expression "middle ground" for the ensuing position which is a wording from Varela's buddhist background which I understand here thus: reality is neither bland illusion, nor brute fact, but poetic transformation.

The core concept of the "Embodied Mind" *sensu* Varela et al. is "reciprocal specification". World and agent are mutually specifying one another. The agent "enacts" a version of herself which is about the world, which is connected to reality, but in a subjective way. Reality is always accessible, but it is so only by being interpreted, put on the stage of subjective dramas, re-invented. This is not an act of absolute authorship, but a co-invention together with others and the whole world.

Varela et al. (1991) derive their idea of "reciprocal specification" partly from research in psychological categorization (Rosch 1978). Their argument is reminiscent of the position held by the most prominent historical representative of embodied phenomenology, Merleau-Ponty (1964). By stressing the common birth of phenomenal world and subject standpoint in perception, Varela et al. try to overcome the danger of solipsism which every strong constructivist, and to some extent, also a biosemiotic position is liable to be subject to (see above). The common genesis of world and living agent provides the key for understanding the way organisms so remarkably fit into their environment, and why communication at all is possible.

Dreaming the Other, Being Dreamt by the World

On an embodied level, the process of perception is not so basically different from the fundamental self-limitation of organism, generating not only self, but also non-self through the encounter with the exterior that has to be excluded from the process of creating an identity. The exterior becomes a pivotal component in the construction of self. And, *viceversa*, self conclusively marks the creation of the phenomenal world.

In this communion the other is reality, not fiction. As necessary part of a common enaction it is even amplified by a kind of absoluteness, as only something material can enter into perception. Varela et al. (1991) try to explain this mutuality in their analysis of human colour perception. They show that colour is created by culture and the underlying neurological structures together. What we perceive as colours is not merely cultural, but neither only biological. It is an interpenetration of both: "In opposition to the objectivist creed colour categories are founded empirically, in opposition to the subjectivist view they belong to our common biological and cultural world. The example of colour thus helps to the insight that [. . .] world and perceiver specify each other" (Varela et al. 1991:237).

This stance is prepared already in Merleau-Ponty. For him, the living body is the arbiter about how the world is perceived. The living body in its existential concern is the interpretant of any encounter with the Umwelt. As Merleau-Ponty observes: "An object is not really *given* in perception, but experienced and internally reconstructed insofar as it belongs to a world, whose basal structures we find in ourselves and of which it represents only one of the possible concretisations". (Merleau-Ponty 1966:377). Both – perceiver and perceived – are in some respect the same, and in another they are incommensurate. Perception has aspects of both.

Reciprocal specification is one way to state that the experience of reality establishes itself as a cascade of mutual transformations in which both sides transform one another through each other. The perceiver becomes what he is through what he perceives which is only enacted by the perception (−creation). This has become crucial for the basic hypothesis of embodied cognition today: Mind is not located in the brain, nor in the whole body, but it is enacted somewhere in between a subject's physiology, her concepts, the world's affordances, culture, and social constraints.

Enactivism: Imagining the Other

In an attempt to push "enaction" to a breakthrough in cognitive sciences, in particular Ezequiel di Paolo, Evan Thompson and Hanna De Jaegher (Di Paolo 2005; Thompson 2007; di Paolo and Thompson 2014; De Jaegher 2015; De Jaegher and Di Paolo 2007) have been crucially refining and reformulating the notion of interbeing, or a co-creation of sense. With this, enaction has become the central focus of description for the phenomenon of life. In its middle stands the notion of unity of "interior" and "exterior" aspects which is so important to the argument in this book: Interior is enacted through the exterior. Mind and Life are coextensive.

Enactivism is an extension of Varela's, Thompson's and Rosch's (1991) primary notion of reciprocal specification, which is that although there might be two spheres (exterior material appearance, interior inward experience), both come to be only through one another. Consequentially, individual identity neither arises as a purely biological individual process, nor as a purely social determination, but as a dance of both. There are always several realities that interpenetrate. The perceptual experience, but also the bodily expression, or the social ambiance in a given community, are the enactment of these transformative forces.

It is not by chance that enactivism derives its name from a poetic practice. It comes from theatrical play. An actor "enacts" the feelings which the character he represents supposedly feels. By writing his lines a poet enacts the energy the world's vibrancy (or the world's deep pain) awakens in him. Therefore, already the term "Enaction", taken seriously, has its own poetics behind it. A perceptual world which is enacted, an identity which is enacted together with and through others, is an activity genuinely expressive of concern and feeling, and it is so by enabling participation.

Enactivism is a deeply poetic metaphor. In this we witness the cognitive sciences opening up to the felt dimension and to inwardness, in the same way genetics had embraced final causality through incorporating the code metaphor. But much work rests to be done. We need to take the poetical force of the term "enaction" seriously. This means that we have to introduce a truly subjective perspective – and not only its description – into biology. Actually Varela was sensitive for this, when in his crucial 1997 paper he pointed to the necessary "surplus" of meaning in cognition, and to the "imaginary dimension" every autopoietic act opens up.

All this is based on the precariousness of living beings, which are not only fragile (always open to decay, hence prompting their self-procucing activity), but even are their own negation. They are what they are not – and this makes them need to enact what they desire to be, through transforming themselves and others, transforming themselves into others, and others into themselves. Di Paolo and Thompson (2014) see this inherent negation and incompatibility in their recent *précis* on enaction. They write: "A precarious, operationally closed system is … always decaying. The "natural tendency" for each constituent process is to stop, a fate the activity of the other processes prevents. The network is constructed on a double negation. The impermanence of each individual process tends to affect the network negatively if sustained unchecked for a sufficient time". We could thus say that matter's tendency to take on the lowest energy level and inwardness' desire to unfold, embrace, entangle and connect, reciprocally specify into the concrete soma of an organism in a concrete given situation and moment.

The Living Self

Enaction means that reality is always in the making. It is not here nor there but at always in between, freshly created. This forecloses any final objectivity. As each contact is transformation, no pure "as such" is findable. We will explore the consequences of this in the subsequent chapter. We will again come back to this in the very last one, where I want to argue that biology in general – and biosemiotics and embodied cognition research in particular – finally have come to the place where physics has found itself at the beginning of the last century. At that time, quantum mechanics and relativity theory showed that there is no absolute objective reality but a huge tangle between observers and observed. Aliveness makes this visible. Actually we would not even require embodied cognitive science to tell us this. We do not need to be taught, because we know by doing. We are enacting inwardness through outside – body – all the time, as we are feeling bodies in connection.

Coming to terms with this entanglement means for biology that we have to say goodbye to making objective statements about lived biological reality, as Sociobiology, Evolutionary Psychology or Ethological Ecology still partially do. The phenomenon of the living is shadowed, it is enclosed in its flesh, to be experienced through participation, and then difficult to put in words, at least in those that are not poetic, and hence another enigma. (Poetry in this respect means to answer to a question with a question). Living beings, the living other, and also the living self remain unfathomable.

Still, this impossibility yields the chance to understanding. The other remains other, but I can participate in her. Di Paolo and De Jaegher (2016), referring to their own previous work (De Jaegher and di Paolo 2007: 504), write: "The experience of the other never achieves full transparency or full opacity, but rather intermittently moves through regions of understanding and familiarity towards provinces of misunderstanding and bemusement, corresponding to periods of coordination or breakdown respectively." Or as John Bryant (2006), writing about Melville's *Moby Dick* expresses it: "The essence of life is contained within the substance of life and cannot be 'extracted' without killing life; full transcendence, therefore, is impossible".

Enaction details this performing, the "mise en scène", the putting-on-stage of freedom through need. In this respect, Hans Jonas with his characterization of the organism's reality as "needful freedom" had already presented the core idea of enactivism. Freedom is transformed through need, the continuous desire to be in connection, and yields a fragile, and sensitive body. In the vein of Italo Calvino we could even say that freedom imagined through needfulness yields a receptive body. And need imagined in freedom yields the "endless forms most beautiful" (Darwin 1859:490) which the earth's history has brought forth.

Needful Freedom Exists Below All Differentiations

Needful freedom transcends all domains. It is not a biologial problem alone, neither is it an intricacy of social identity. It rather characterizes being-in-the-world as such. It is existential. It is the situation of the core self. It is the condition within poetic space.

This "core self effect" opens poetic space. It makes bodies vulnerable to social chill as if it were a frostburn and it leads to inflamed tissues as if they were infected with a pathogen. But this core self effect also makes the soul open to a verbal caress as sore skin is to a cooling touch. We exist within poetic space, and this is the source of the magic of social exchange as it is the source of the self-construction of the body.

Di Paolo and de Jaegher (2016) in a very delicate way observes this when they quote examples which show to what degree non-bodily processes can effect the wellbeing and functioning of the body. A bad relationship can make one sick, and a tender one has empirically proven healing powers: "Kind marital relationships can make a spouse's wounds heal faster, while hostile relations can slow down their healing" (quoting Kiecolt-Glaser et al. 2005; Gouin et al. 2010). The other is not entirely outside, but takes part in the processes of the own self-construction.

We could even say: Two people – or people and nonhuman beings – can share the same core self, or are transforming one another at the core self level.[1] But that still does not mean that people meet outside their bodies in the sphere of social identity. It only means that there is an existential dimension which lies before the differentiation into domains or disciplines.[2]

Meeting at core self level means to meet within the primordial polarity of "need and freedom". Here we can see that the difficulties of being-in-connection, so crucial for a living being, arise through the same deep incompatibility which characterises also the individual identity. The primordial tension in any relationship and not only in a human one therefore is not about self vs. other. It is about the problematics of polarity in relationships, the fact that they only can be construed as a negation of polarity which still needs to respect polaritites. This intricacy ultimately touches the constitution of the subject as intersubject. This is the primordial tension of being a subject in the world and as world. "The primordial tension is more subtle and pervasive; it is in place even if the other is not present as an other, but as relational patterns that affect my sense-making and are affected by it" (di Paolo and de Jaegher 2016). The problem of the other is the problem of relatedness as such. It is therefore not different from the problem of self. The problem of relating is the problem of self.

After the sun has set in Sant'Andrea off Elba, the thunderstorm, gray and violet, has moved on toward Corsica. The sea simultaneously mirrors the colors of the atmosphere and shakes them off, while its choppy suit of armor takes on every hue: turquoise, sky blue, gray, orange, violet, ultramarine. The ocean has no colors, it has nothing but energy. The ocean is the "wine-dark sea" that Homer celebrated in song. The sea has not the colour of wine but through the light it is emitting shines the drink's ages-old transformative force, the power that enables the actualization of living things. It is a power that makes things more real and that lends itself to everyone who carries it further and transforms it. It is not a physical quality nor a metaphor, but a common space of being which radiates out aliveness.

---

[1]Comparable to Marshall Roesenberg's (2003) "heart level".

[2]It adds some flesh to these philosophical deliberations to read de Jaegher's (2015) phenomenology of the loving relationship in nuce: "On the backdrop of this analysis, we also get a new perspective on lovesickness, a paradoxical feeling familiar to many of us. It seems a simultaneous enjoyment and suffering of one own's life, one's living self-affection as modulated, transformed, moved, upturned by another in the ebbs and flows of an intricate and intimate real-life encounter. It thus hints at the simultaneous existence of solitude and inter-affection, both equally existential." (p. 127).

To pay justice to reality, we need an attitude of inclusivity, of mutual acceptance between attitudes, bodies, identities and sensations. We need the affirmation of belonging and a willingness to engage in an ongoing negotiation within a reality that we recognize as a commons of perception. We can adapt our behaviors to its ongoing transformations and amalgamation. Since this is the inescapable existential reality of life itself, we must acknowledge that the fertile wild ultimately cannot be denied, suppressed or enclosed without a profound constriction on our own freedom.

Chapter 9
Flesh

> *"D'autres sont partis à la recherche de l'étroit passage du Nord-Ouest qui permettrait de faire communiquer les deux continents, sans songer que les deux continents sont contenus les uns dans l'autre . . . "*
>
> Edgar Morin (2001)

I want to complete the discussion of being-through-the other with the thoughts of Edgar Morin, a French philosopher who in his work has touched upon many of the major arguments which the enactivist view unfolds. Morin started his work in the 1950s. He elaborated on some of the core ideas of biopoetics well before they had been explored by embodied cognitive sciences.

Even today, embodied cognition research does not take Morin's work regularly into account, although Morin substantially reflects about the phenomenon of life. Even Varela did not pay much attention to the philosopher, although he was researching and teaching in the same city, the French capital. Morin sketches a view which prioritizes the poetic dimension as the all-encompassing centre of embodied experience. "The poetic notion is no epiphenomenon", he writes (2001:127). It is rather the key to relate to organism. For Morin, poetic understanding and experience are the gate through which we can understand the living.

I do not intend to extensively review Morin's work, but I want to shine some light on a couple of strikingly coinciding thought figures, which can reinforce some of the arguments developed in this book. Morin has developed a particularly high sensitivity for the paradoxical nature of all life. In respect to contradiction inherent in biological processes, and their consequences for a picture of life, and of our role in it, we can learn from Morin's work and use it to gather further insight. Therefore, it might prove fruitful to enlarge the current knowledge body of enactivism by Morin's more fundamental visions. In particular, Morin has build an entire philosophical worldview on the idea that "the other is an inner necessity" (2001:68) for self.

A. Weber, *Biopoetics*, Biosemiotics 14, DOI 10.1007/978-94-024-0832-4_9

Dialogics: The Other as Inner Necessity

Morin stresses the inseparability of self from other. This attitude makes his standpoint so beneficial for our task of a poetic philosophy of organism with its scope of integrating two sides which stand in mutual contradiction and therefore generate the temptation to flatten out either one of them. Morin emphasizes strongly that we must not renounce of either side. An individual exists only through other, in all precariousness ensuing from that entanglement, but also in the absoluteness of being a single center of concern. Morin does not attempt any reduction but rather urges to develop this antagonistic constellation into an ontology of paradoxical existence. With this, he clearly steps out of traditional western logics. But he firmly remains on empirical ground, in the same way that autopoiesis research and embodied cognitive science explore what the body really does and only from here come to their philosophical conclusions.

We can see that double absoluteness of two contradictory perspectives when Morin (2001:69) states that "a subject structures itself through the mediation of other subjects before even knowing them . . . the subject emerges from the world by integrating itself into intersubjectivity. Intersubjectivity is the tissue through which subjectivity can exist, its milieu of the subject's existence, without which it perishes. But . . . the subject does not dissolve within intersubjectivity which nonetheless assures its fullness. The subject's drama is to be not only a transmission relay in a tissue of intersubjectivity. It retains its irreducible self-affirmation." Here truly two continents are contained in one another.

Logically, this is the relation of a set being its own element, an operation a binary computer will refuse to do. This contradictory relation cannot be done on one level of description alone. You need two levels of scaffolding for it, which already must contradict one another and must not be reducible to one another. A contradictory relation therefore transcends pure materiality. Causal reactions must be free of contradiction and cannot be described by a second level of contradicting causal relations. The logics of contradiction therefore dictates that we need another level of reference, which is outside the actual physical space, as is the "code" of DNA in regard to the soma, or a perspective of concern in regard to the autopoietic organization, or language in regard to reality.

In the realm of organisms, binary logic is inadequate. Living reality, Morin (2001) claims, should be described by a "dialogics" rather than a binary logics. Morin's "dialogics" and Varela's "reciprocal specification" have much in common. They yield a relationship which is dependend on the situation, and even more on the interaction of two (or more) participants, and which is rather a collective imagination than an objective protocol. A dialogics is imaginary but still contains all constituents, transformed through one another.

Morin (2001:281) defines this transformative interpenetration as "complex unity between two logics, entities or instances who are complementary, competitive or antagonistic, and which nourish and complete each other, but which also oppose and fight one another". Morin's dialogics does not try to eliminate contradictions but explicitly seizes on them to illuminate the point. Living reality is a logics of dialogue and polyphony, of encounters, conversations, mutual transformations and interpretations, a logics of negotiation and striking compromises (Morin 2001:272). This kind of logics is also inevitably dynamical. Dialogics is built on constant renegotiations of the relationship (as is the body built on a constant production of self from the surrounding matter). It is a process, not an outcome.

Subjectivity Beyond Embodiment

We can exchange messages about what it is to be alive through the eloquence of our fragile bodies. Social interaction which includes bodily perception and expression, speech and touch, is a deep part of social cognition. This exchange never takes place between individuals that are complete for themselves and closed to the outside. As we have seen, a subject is not closed in a material sense. But also in an experiential or symbolic sense it is constantly open.

The subjective inside therefore can only be understood as already-other, as "interbeing". Only through exchange with the other a subject can arise. The dialectics between self and other is prefigured by the fact that the autopoietic circle of self-affirmation is possible only through the constant metabolic flux of "other" through the centre of self. Matter changes, and only by this change, by this continuous "sucking in" of the world and the subsequent transformation of this "other" into "self" does an individual arise. This holds true for cells, individuals, and groups of individuals. For humans to become a self, the social other is needed. But the social other is not structurally different to the level of existential identity. It enters into it as a part of the constitution of self, just as body processes do in their "vital import".

As di Paolo and de Jaegher (2016) state, "Social interaction processes enter into the core of our self-constitution. In and through social interaction, we can truly affect each other, even each other's self-maintenance and self-affection". The particular symbolic importance of the other in constructing self has become obvious in recent psychological research concerning newborns. Only a couple of decades ago, infants were treated as blank slates existing in a mere vegetative state (even the time when infants underwent surgery without anaesthesia ended only some decades ago).

To reasses this fundamental scientific belief, the work of developmental psychologists Andrew Meltzoff and Keith Moore has been crucial. Their research (Meltzoff 2005) paints a picture completely different from the idea of babies as not-yet-full-beings: it describes newborns as wholly developed subjects.

Meltzoff observes that infants are able to imitate a wide range of facial gestures immediately after birth. For Meltzoff, this capability to see the other as a body and to echo it with one's own says a lot about the relation between inside and outside: "The ability of young infants to interpret the bodily acts of others in terms of their own acts and experiences gives them a tool for cracking the problem of other minds."

Babies can understand the gestures of others, because through their own bodies, they are supplied with a model for "mind".

During the last two decades Meltzoff and his co-workers have shown that the ability of newborn babies to imitate faces is not a reflex, as it has been shown to be too variable, and too good an imitation. Apparently, neonates must know, or rather feel, that what they see in the face of another human being is what they can produce on their own face. To be able to imitate, they must make a direct relation from their inside to their outside. They must know that they are outsides with an inside, as well as that the other is this too. Meltzoff and Moore (1995: 53–4) have suggested as an explanation a primordial connection between self and other within each other's core self, which they call a "supramodal framework".

Through the core self, which is body, feeling, and expression at once, infants are able to feel "an intrinsic relatedness between the seen bodily acts of others and the internal states of oneself... This primitive self-representation of the body may be the earliest progenitor of being able to take perspective on oneself, to treat oneself as an object of thought." (1995: 53–4). By a related argument, primate researcher and cognitive philosopher Frans de Waal puts forward a universal "perception-action-mechanism" within and between all higher animals, making subjective states of the other accessible as experiences of one's own body (see Olbrich 2009: 119).

Encounters reach the subject not as *qualia*, but as *values*. They are experienced as meanings in relation to the organism's existence. They are, as Cassirer's expressive qualities, basically values about good or bad. They are vital import to the core self, supramodal, or synaesthetic, as they are interpreted holistically according to their meaning for the ongoing realization of the living (Weber 2003).

Subjectivity arises from a primordial intersubjectivity. To the newborn child, the comprehension that outsides have insides follows from the very fact that it exists: subjectivity is given to him as inwardness with an outside. We can infer that for the newborn there is nothing more normal than the fact that his mother is a subject with a body and an emotional inside. Any subjectivity conceives of the world according to the experience of this primary relationship between inside and outside. For a living subject the idea that outsides have intentional insides simply is the normal viewpoint. A subject experiences itself as something which *is* an inside with an outside. For this reason a child experiences the whole world as animated.

Infants experience themselves as individuals who are able to reach out and connect. They do not experience themselves in a continuity with other adults, and particularly their caregivers. There is no "primary narcissism", although this is still believed by the majority of developmental psychologists and psychotherapists. Many psychological theories of personal development and psychopathology still circle around the idea that a child cannot distinguish between itself and the other world, in particular its mother. The myth of prenatal "oceanic feeling" became an unproven standard in infant psychology through Freud's teachings. But it is probably not the case that an infant does not feel any separation between itself and the world. To experience oneself as subject means to experience the self as autonomous, longing for connection.

To feel one's own separation from the world is the necessary precondition of subjectivity. It is the "subtle tension" of existence which is showing through the concern of core self. To be an embodied subject means to be separated. The subject needs to be able to overcome this separation from moment to moment by building connections and through them imagining oneself anew through other. A healthy development of subjective identity therefore needs the other, which "sees" the subject through transforming her feelings by responding to them. This gentle response is not a mirroring, however (as is also believed by many developmental psychologists), but an imaginative – and therefore consoling – transformation of the infant's feelings through the reaction of the caregiver, who acknowledges the feelings but embeds them in a wider context of care and connection.

Subject as Inter-Subject

The sphere in which mutual reciprocity yields perception, in which poetic expression is not an arbitrary judgement by an individualistic mind but the objective imagination of reality, has been described by the French philosopher Maurice Merleau-Ponty as the "flesh of the world". After the discussion of the subsequent pages we can see more clearly what Merleau-Ponty was aiming at.

The "flesh of the world" is not identical with the physiological tissue of the living body, but it is also nothing without it. It is the sphere of existential reciprocity which brings forth living reality. It is characterized by the fact that experience is never "about" something, but always happening between one being and another through mutual transformation. Only according to the degree to which the experience is shared by both, it can be real. Only according to the degree it is a real transformation, it does exist. As Shierry Weber Nicholson (2003) observes, here "interior and exterior are the reciprocals of one another". What we conventionally see as a matter of touch or vision alone – a world "given" to our senses, the test tube environment "given" to a bacterial cell, a thing sitting before our eyes or a sound coming to our ears – is a matter of reciprocity, of mutual creation, which makes both partners more real.

Merleau-Ponty calls the ensuing mutuality, in which perceiver and perceived become entangled, "chiasm". Only through chiasm, in the middle of both, there is reality. The "body is flesh as the world is flesh, senseable, see-able" (Weber Nicholson 2003). Flesh in the sense of Merleau-Ponty is not the meaty substance, not protein and tissue, but an existential sensitivity, a felt meaning and a poetic expression, which comes into being through participation. This participation, however, is assumed by feeling bodies. The "flesh of the world" to some degrees is the flesh of the body, but it can also be there as verbal caress which stirs the soul as a soft wind stirs the fine hairs on the skin. The law of the flesh of the world obeys poetic objectivity.

Merleau-Ponty underlines the fact that perceiver and perceived are not to be separated by a dualist conception. Rather, as the perceiver is an embodied subject, his body mediates between world and self. We can see Merleau-Ponty's work as a research into the embodied conditions of subjectivity. And not only in his first phase, finishing with the *Phenomenology of Perception*, Merleau-Ponty's ideas show striking similarities with the biosemiotic attitude found also in Cassirer.

Like in Cassirer, it is the embodied subject who unfolds the dimension of meaning and hence produces a primary semiotics: "Already the mere existence of living beings transforms the physical world, makes here appear 'food', there 'shelter', and lends stimuli a meaning which they had not by their own reason", observes Merleau-Ponty (1966: 23, my translation, AW).

Still, Merleau-Ponty stresses the relatedness and the common nature of perceiver and percieved more than his German colleague. In his later works, Merleau-Ponty's emphasis shifts even more to that reciprocal view. His approach changes from phenomenal psychology to ontology of subjectivity. Rather than the act of perception, subjectivity itself becomes the philosopher's focus. Merleau-Ponty insists that it is the specific human building plan that shapes our world. But at the same time body does not determine a particular Umwelt. It rather lends itself to multiple dialectics of inside and outside, leading to ever higher levels of symbolical interaction.

In his late writings, Merleau-Ponty expands this notion of dialectics into an ontology of reversibility in the field of perception. Subjectivity is only possible as intersubjectivity, because for perceiving in the first place, for experiencing meaningful signs, the subject is dependent on the other. The subject is, at its very constitution, only possible as incorporation of the other. This ontology is still related to the embodied origin of subjectivity. Its possibilities are inscribed into the body. But body is already a complicated dialectics of inside-outside-relationships. It is the paradigm of inside-as-outside.

Thus, for Merleau-Ponty, the paradox of immanence and transcendence is a paradox of life itself. An absolute dichotomy between self and other is *not* possible, because self is only possible as an entanglement between self and other (cf. Reynolds 2004). Self and other are intertwined, but they are so by being at the same time divergent. They keep an 'intimate distance', to paraphrase a title of one of the last papers of late Francisco Varela (2001). This discussion is particularly important regarding the debate about an ethics of the "absolute other" favoured by thinkers as Lévinas and, in his footsteps, Derrida. Reynolds (2004) discusses this point and shows that by conceiving of a radically different other, as Lévinas and Derrida sometimes seem to intend, we are confronted by the problem of solipsism 'through the backdoor'.

Things are very different, however, when other is seen as a part of self, which is not only absolute in its existential concern, but at the same time only possible through a primary intersubjctivity, in which other is already built in as an absence, a needful lack to be compensated for through the surplus of imagination. It seems that only a middle way of 'intimate distance', where the position is always to be negotiated, brings us into the position to integrate intrinsic values of subjects and values and goals of the other.

To Touch Is to Be Touched

To better understand Merleau-Ponty's argument we can take a look at his famous analysis of the human ability to touch as being touched, which stands as a model for the whole perceptual encroachment of other upon self. If I touch myself, I can feel myself at the same time from an inside, as being touched, and from an outside, as touching me. If I touch you, I feel myself being touched by you as well. In touching you, I feel that you are touched. Every touching means to *be* touched, and, as Merleau-Ponty concludes, even every seeing might mean to be seen. Perception is a reciprocal process. Perception is a meaningful space in between.

The point is that 'I am experiencing myself in a manner that anticipates both the way in which an other would experience me and the way in which I would experience an other' (Zahavi 2001:161). "Our embodied subjectivity is never purely located in either our tangibility or in our touching, but at their intersection and where the two lines of a chiasm intertwine and cross", as Reynolds (2004) observes. This 'flesh' is the region where meaning is generated from the intertwining of embodied interests in perceiving outsides and hence creating subjects. The flesh is not biochemically defined as the biosphere. It is rather the semiosphere, which includes the physical biosphere, the dimension of reality where meaningful encounters are formed.

Subjectivity in its embodied dimension is always given as an inside and an outside. It is our body—and for any other embodied being it is her body as well—that inevitably throws the subject into this double view of inside and outside. Intersubjectivity hence is 'rooted *a priori* in the very relation between subjectivity and world' (Zahavi 2001: 165). And, the other way round, subjectivity is already intersubjectivity, as it is a mediation between inside and outside from the beginning, where one can only be given as expressed, or symbolised, by the other. Outside, or other, is the symbol of an inside. Inside is the meaning of outside, or the other.

"Inside" is inwardness. It does not exist on a physical level, but on an experiential level. It is not what is "inside" a box. Inside is what has meaning for a concerned being, hence creating his standpoint of concern. Inside is not the object of introspection, but it is signification. It is 'inside' as the domain of relevance included into what counts for the system, but not necessarily as 'in' the system.

Merleau-Ponty's position has been empirically strengthened by Meltzoff's and his co-workers' experiments discussed above. To be able to imitate, infants must make a direct relation from their inside to their outside. They must know that they are outsides with an inside, as well as that the other is this, too.

From this discussion we can elaborate a definition of subjectivity which goes further than the idea embodied cognitive science generally holds of the embodied subject, which is sometimes seen as being "given" before true intersubjectivity becomes possible. To my eyes, however, the subject itself is primarily intersubjective. This is what the infant in Meltzoff's experiments showed. Knowing itself as subject meant knowing itself through intersubjectivity. Being a Subject means understanding that one exists through the other, as self is already the experience of other.

The level on which self experiences itself as already through the other is not "the body", however, nor "the society" (and its intersubjective interactions). It is the sphere of meaning in which other is existentially relevant to self, and which is "inside" body, as inwardness, or core self, and cannot be separated from it. In this sense, subjectivity-as-interbeing is also always self-as-world. In being itself, as through-other, a subject is the world, as through-self. We can therefore add three more principles to our catalogue of poetic worldmaking. Although I have touched upon them in the above discussions, let me here sum them up again:

1. Subjectivity is always intersubjectivity, self is what it is not.
2. Subjectivity/intersubjectivcity is always embodied.
3. Intersubjectivity is not body. It happens in a poetic space, which is beyond body, although including it, but doing so from the perspective of its existential imaginary dimension.

The knowledge body of ecology is a proof for the middle ground, the chiasmic interdependency of all subjects and things as well. Its paradigm is hypercomplexity, the entanglement of everything with everything. Here we have real interdependencies which make only possible the diversity of its middle ground. As authors as Lynn Margulis have shown, the species diversity of, say, a red sea coral reef, which awaits still full explanation in terms of the neodarwinian paradigm, is only possible as a network of infinite interactions. If we cut this network and remove some of its parts, it will change as a whole.

The evolution of individual species, a snail feeding on algae for instance, cannot be explained by the outcome of external factors alone but only as a feature of the gestalt of the whole evolving reef. We see here clearly that a phenomenon, an embodied reality in space and time, is only possible through the interaction of innumerable forces and fates. We have not only one level of meaning, but also an example for the reality of the 'flesh': it is, in case of the coral reef not only reciprocal relation, but also swelling body, the most delicate organic layer which covers the stone of the coral heads.

Being the Other

While self already is an interplay of self and other, a relationship between an individual and another involves an additional level of paradox which is about incorporating this physical other into the physical self. Subjectivity is intersubjectivity, and the relationship to another individual follows the same dialectical laws of establishing identity through separation and is not profoundly different from the way an organism self-creates in an autopoietic fashion. The ensuing relationship is a process of mutual trasformation which deeply enters into the constitution of the participating subjects. The other is inside of me.

Biology has explored this through the discovery of mirror systems in the brains of various species, including our own (Rizzolatti et al. 1996). Formerly called "mirror neurons", it became quickly clear that rather than single nerve cells whole centers of the brain participate in establishing a connection to the other which runs below conscious experience and beyond the physical separation of two beings. Mirror systems hardwire one organism to another, although both are physically separate. Through this, empathy is generated. This is as inevitable as a feeling response, as is the own body's pain. Empathy is generated as a resonance frequency of the self's own sensitive body.

Mirror neurons connect a subject's perception of other bodies to her self-perception so that she perceives another body's gestures as though they were her own. There is no choice in doing this: Mirror neurons are connected to the brains circuit as if they were conveying information about the own body. They are the other's body in my own, forcing me to feel his or joy as my own, as if we were one bigger body. It is still not clear if the mirror system is a specific functional sub-system of the brain or if it is simple the ways vertebrate brains work in the first place, exemplifiying the default way a living being's central nervous system reacts to other beings. Maybe all brains are mirror systems.

Through the mirror systems the borders between two individuals become permeable. If the self can only participate in the experiences of another being by feeling them in his own body, and then as his own core self, self and other have swapped bodies — or can exist only as organs of one huge body, which extends through many individuals. Through the mirror system, we can imagine all organisms connected in a meshwork of experiences, which are existentially real even *before* they have obtained their explicit shape in individual bodies.

We can see that the mirror system is a variant to the way the core self at the same time *is* the physical body and totally detached from any physical location. Core self and physical self are "intimately distant". By feeling through our own body the existential meaning another organism undergoes, the other enters into our own core self, although being physically separate and distant. The core self does not need a physical container. It can be enlarged to the whole world. It is not contained inside the body. By this, identity, and mind, are extended into the world, albeit not in a spatial sense (Clark 2008).

Babiloni and Astolfi (2014) showed that brains of two individuals even synchronize when one's presence is only imagined. They discovered a deep entanglement of brain and interactional dynamics of two human subjects through embodied dynamics even if one of those persons presented to the research subject through images. Beyond visual representations, probably mere imagination suffices to have the same effect. Dumas et al. (2010) could show that in the presence of the other this interbrain synchronization occurs faster than bodily movements, the so-called "body language", from which it miraculously seemed detached.

Di Paolo and de Jaegher (2016) ask accordingly: "How is it that an interactive pattern appears to affect the oscillation phase of neural groups occurring in two distinct brains at frequencies more than one order of magnitude faster than the interactive movements?" They assume that a possible answer to these puzzles is "that interaction patterns produce an entanglement between the brains of the participants", and that "the processes "at the top of the hierarchy" are the emergent rhythms of social interaction" (di Paolo and Jaeger 2016). The rhythms of social interaction however are not material entities. They are, as the authors say, rhythms, hence patterns which exerce a poetic meaning, like rhythms in music or speech. They are manifestations of a shared core self, and these manifestations present themselves through poetic expression.

This poetic connection its hardwired. The mirror systems show that we cannot avoid being the others, and feeling in their bodies. Because mirror systems are not unique to human brains, this is valid also for other animals. There are gestures – tender and menacing – which signify existential constants which an organism can understand to the degree it is able to embody these gestures. To make them possible, organisms need not to be closely related. We feel the pain of suffering invertebrates and even protists although their bodies are considerably different from our own. But although a crushed cockroach has six legs and a relatively hard shell, its basic body gestures are the same in regard of how fee feel them within our own existential concern. The skin of our back also constrics at the thought of an insect writhing in pain.

We cannot walk through the world without experiencing these gestures in a sort of absoluteness. They roll right through us. The towering oak, imposing its resilience on our body, the swiftness of the hawk, the sweet tingling rivers – we can feel all these gestures as possibilities of our own bodies even without having understood anything about them. These connections are made without our consent, and they carry us away. It is still like the californian environmental pioneer and nature writer John Muir (1979) observed when he described his reaction to the "blossoming waters" of the hills and valleys north of San Francisco, which have been conserved in some small places, like in the bear valley south of Point Reyes Station. Muir felt that the "rivers flow not past, but through us. Thrilling, tingling, vibrating every fiber and cell of the substance of our bodies, making them glide and sing".

Poetic Matter

What is alive resists any and all synthesis. Inner, immaterial and experiential identity, coiled within a material body, is itself the greatest paradox. This identity, which becomes real only through a body, has no separate physical mass and occupies no space. And yet it profoundly alters the physical world and space through a continuous self-referential process.

Any exchange – of goods (in the economy), of meanings (in communication), of energy (in metabolism), of identities (in the bond between subjects) – always has two sides: an external, material side, in which stuff is relocated, and an internal, existential side, in which meaning is expressed and experienced. Reality is creative and expressive precisely because it never lets itself be reduced to one of these sides.

Since all processes are founded upon relationships that convey meanings (which all subjects experience as emotions), the most appropriate way to formulate such a reality is through the idea of *poetics*. But this poetics is not made from thin air – or abstract thought. It is found between dense bodies.

Poetics is at the same time symbolic and material and therefore it is inextricably linked to social communication, exchange and interactions with others and the environment. Poetics describes the world that we experience in the perspective of the first person – the world in which we are at home in an intimate way and the world that we seek to protect through political arrangements. This first person of experience is always real in flesh and blood – a thing as well as a standpoint. Economic exchange, which is a meaningful householding among living beings, can also be described as a poetic reality.

Perhaps one could call such a perspective *poetic materialism*. Any cognition through relationships can take place only in form of poetic acts. Living relationships, however, organize themselves among bodies which constantly transform themselves, which grow and decay. In systems in which change occurs dynamically as participants seek to negotiate and transform each other, experience cannot be expressed as a fixed identity, but only as the transient expression of one through the other – in other words, poetically.

The poetic dimension is simultaneously a modification of the individual and a modification of the whole. It becomes distinct and visible in an individual only through forms of experience and symbolic expression. We are transient transformations in a larger process defined by material/semiotic relationships. Viewed in this light, reality is revealed as a commons of those perceiving and those perceived, and their ongoing interactions. Its objectivity is not simply an academic discourse. Nor is it invented or constructed by human culture. Instead, reality is both a way of describing the world as it is and a set of experiential practices. Like Aristotle's ethical ideal of a mediation between the "wise and the many" (Nussbaum 1986), the ontology of the world is never fixed and unequivocal; it is always process, always birth, always becoming. The goal lies in participating in the enterprise of creative aliveness in order to make the world more real.

Chapter 10
Objectivity

L'acte poétique est un élément de connaissance du réel.

(Edouard Glissant (1996))

In the last section we have seen that an organism's worldmaking cannot yield an objective view, a representation of reality. The world where an organism lives, the organism itself, and its fellow beings, are to a large degree imaginary, the results of a commons of imagination. This imagination is a material process. Its imaginary dimension is not arbitrary. It carries elements of the self and of the world in mutual transformation. We have seen throughout the argument of this book that the imaginary dimension precisely is there because living beings are bodies which have needs stemming from the intrinsic contradiction of being an individual through matter. The freedom of the imaginary is thus due to the necessity of embodied needs – and to the "confusion", as Kull puts it (pers. comm.), the organism faces in trying to provide for these individual needs through connection with other.

Only this imaginary specification is able to bring the mutual sides into contact. Still, there is no reason to be nihilistic (as has been so fashionable in western culture and science) about the fact that an objective view is never attainable. The substitute for a thinking which worships substances – like "objectivity" – can not be a standpoint which gives up any firm ground, but knows that firm ground is always as firm as it is imagined. This thinking is the way the living body inscribes itself into the world. Its objectivity is always relative to embodied needs.

Organism defines a particular kind of objectivity. It outlines objectivity needed in a realm of flesh and blood, suffering and desire. It is not the fictional objectivity which still dominates much of today's thinking, which presupposes a fixed, static reality which can be known, and to which organic machines more or less efficiently adapt. This is the objectivity of separation. It is the objectivity which still rules. Organism, however, in its way of always searching relations which reciprocally specify the participants in them, creates an objectivity of connection. And even more: As we are organisms ourselves, we know about this objectivity of connection from the inside. We use it all the time. We know how it feels – as it reveals itself not by measurement, but by feeling experience. Imagining ourselves objectively means to feel in contact with ourselves.

If we use an objectivity of connection in order to talk about organisms and the world which is continuously giving life to organisms, we will not only be able to

A. Weber, *Biopoetics*, Biosemiotics 14, DOI 10.1007/978-94-024-0832-4_10

come up with a lot of fixed entities (as anatomical or physiological details, which naturally remain valid as they determine the possibilities of connection). We will rather evoke a world of first person experiences – felt connections. But this is not "subjective only" in the conventional sense – open to games of arbitrary and disembodied discourse. The experience of felt connection is imaginary, though not arbitrary. It is objective, as it is about bodies and their way to enter into mutual interpenetrations. It is the objectivity of the feeling self which we share with all other beings.

As this other form of biological objectivity comes into being through relations in an enactive way, through mutual transformation, we can say that it is poetic. It is poetic in the sense that it is an objectivity which is about being in relation, which expresses the real conditions of all its participants, but does so in an imaginary way, by expressing them through one another.

Objectivity Refuses Purifications

The poetic dimension is the world of our feelings, of our social bonds and of everything else that we experience as significant and meaningful. The poetic sphere therefore is part and parcel of our everyday world of social communication, exchanges, and interactions. It is the world of first-person-perspective, which is always there, and which is always felt and experienced. It is the world that we live in most intimately, and it is ultimately the world for which we conceive and make various policies. The world of economic exchange, which is also social exchange between living beings, takes place in this world as well.

The standpoint of poetic objectivity does not propose an individualistic or solipsistic worldview. Rather, I argue that the subjective perspective of embodied beings is a necessary complement of the prevailing objective approach. Here, too, we must come to terms with the reality of incompatibility – or paradox – in everyday life. As living organisms we have to learn to experience and to describe the world "from the inside" (emotionally, subjectively, socially) while also treat it as an external physical reality that exists "outside" of us.

Bruno Latour has argued that any procedure that attempts to "purify" the biosphere by insisting upon its physical dimensions only – while denying that it is a sphere of meaning or "semiosphere" as well – will generate even greater, albeit hidden, tensions. The sorting of reality into the categories provided by a dualist worldview is a case of emotional repression. And just like psychological repression of inner antagonisms it will only generate neurosis. This neurosis of purification is inscribed in our present culture (Latour 1993).

If we rely on the shared objectivity of our embodied feelings, we can see the world we know with entirely new eyes. Think for example of an embodied being (a human, an animal, a plant) who is sick, looks sick, and behaves in a way that shows her affliction, although no "organic cause" can be found. In the "old" objective way, this plant, animal or human is declared to be healthy "in reality". It better should stop making such fuss. In a perspective which considers poetic objectivity, this subject looks and feels afflicted in an almost contagious way. Her condition becomes real because the others with whom she enters into connection are affected by it and mirror this affection. Affection becomes an objective process, although no one might consciously feel it. Think of the ouzel, John Muir's "hummingbird of the California water-falls", who even in the harshest winter dives under stones to search for food, his white breast and stone grey plumage echoing the swirling swiftness and rock-hard necessity of its habitat, which thus can be appreciated by another observer.

We could even find a name for this poetic objectivity: We could call it the experience and the expresion of "aliveness" (Weber 2013, 2016). Aliveness is about *being* in a transformative relationship and allowing for future imaginary acts through this aliveness. Aliveness is not reserved for one domain, the personal or the general. Aliveness is not limited to "nature" or "culture." It is intrinsic to all social and biological systems. It has an objective, empirical substance and a subjective, tangible dimension, and it is always interweaving dimensions of matter with perception and experience.

Seen from this perspective, it becomes essential to adopt a *first person* viewpoint as a counterpoint to the purportedly scientific perspective of "objective reality," which is typically expressed in the *third person* (Weber and Kurt 2015). In the ontological reality that we are describing here, the first person perspective is both poetic (rich with meaning, feeling and implications for identity) while genuinely objective (material, scientifically measurable). The first person viewpoint mediates our perception with our material reality, which is only possible from the perspective of a meaning-making self (see Weber and Varela 2002 for more details). Internal, first-person insights that were ruled out by a worldview that accepts only the empirical/objective point of view – because they are not "real" in the material, physical sense – suddenly become valid.

Once natural ecosystems are seen as creatively alive, it becomes necessary to complement rational thinking and empirical observation with the "empirical subjectivity" of living things, and its complement, the "poetic objectivity" of meaningful experiences. This new standpoint cannot be dismissed as a soft, vague emotion, but must be heeded as a critical genre of evolutionary intelligence. As living organisms, we can learn to experience and describe the world "from the inside" (emotionally, subjectively, socially) while at the same time treating it as a physical reality outside of us. Poetic objectivity is a solution to the destructive dualities that since the Enlightenment have isolated the human species as above and apart from "nature." Poetic objectivity represents the missing first-person-centered perspective in human culture that must act as a complement to the dominant but "empirically-objectivist" approach.

Poetic Practice: Sharing the *Conditio Vitae* Between All Beings

If we rely on the finding that we all share lived experiences that are not hidden from the mind but rather constitute its foundation, then connecting on a deep level seems possible. The common, embodied "ground of being" by this becomes the prerequisite for communication. The major obstacle to put this into practice is the fact that humans have too little access to their embodied needs.[1] These needs are nothing other than the species-specific manifestations of our existential necessities as organic beings. They are individual, but also to a very high degree shared. We all need bonding, food, shelter, health, and freedom – not only humans but also animals.

The real challenge to such a vision of science, however, is to learn how we might systematically approach this kind of felt experience and generalize it into knowledge and practice. Such an approach will require not only theoretical shifts but also practical changes. The goal must be to re-embody thinking and re-connect it with the corporal-meaningful rationality of our body-mind and of all other living systems which have been successfully dwelling on our planet for some 6 billion years.

Poetic Objectivity is about trying to establish a logic of sentience beyond the limited logic of "objective" reason. It relies on another type of verifiable objectivity – the logic of our shared experiences as living beings. Pain and joy are objective facts for all living beings because they all feel them. Living agency is an objective fact that unites and transcends all disciplines. The idea of poetic objectivity means that we can – and that we even have to – find a complement of lived practice for any theory – in biology, ecology, economy, sociology, psychology, physics and also the arts. This lived practice might be able to provide a basis for generalizable principles and transdisciplinary inquiry.

But what guideline can lead us to find a suitable lived practice for this new type of science? Does a model already exist, in any spiritual tradition, in the arts, in science? And how can we justify a universal validity for the practical framework once we have found it? A serious consideration of lived practice forces us to reckon with a stubborn, objective reality: Living beings are those natural "facts" that produce value and meaning from within. They are manifest in their desire to stay alive and unfold. Any living being intentionally – if mostly unconsciously – strives to exist, grow, give and receive. The (original) theoretical premises of biological sciences should not allow such observable realities to be marginalized or ignored.

Poetic objectivity can re-integrate individuals on both a corporal and an existential level. It goes beyond an abstract objectivity of the mind to embrace, as well, an objectivity of the living organism. It refers to our shared condition of embodied beings which we could call the *conditio vitae*. Poetic objectivity is possible because of the empirical subjectivity we have discussed in the previous chapters. Being a

[1]This is also the basic assumption in "Nonviolent Communication", Rosenberg (2003).

body is an irreducible fact *and* an experience – as opposed to "*having* a body," which implies that our body is an "other," separate from the self, a thing, a machine serving an immaterial mind in the sense of the "Cogito" introduced by Descartes asserting that we can only be sure about our mind.

Existential Sculptures

But it is possible to claim a subjective, first-person certainty about our body and experience to which in reality even Descartes' famous phrase can be traced back. Being a body and having feelings and socially expressed, nonverbal interactions, are empirical facts. They are also dimensions of living existence that are shared with all other animate beings. Poetic objectivity is about this subjective core self: the existential meaning that any organic being produces from that center of concern that is its self.

The crucial point is that we all – all of us living beings, from the most modest bacterial cell in our guts to you, dear reader – share the experience of a meaningful core self that is concerned with what happens to it and strives to keep itself alive. We all share the experience of imagining ourselves anew by being alienated in our deep core, by running into a multitude of incompatibilities, from metabolism to social relations. As living beings, we all have a genuine and felt interest in continuing to live, and we know the joy and light-footed exuberance of just being.

Poetic objectivity seeks to understand how expressiveness-in-our-body feels and can be communicated, and elaborated upon. The proof of viability of a statement which is poetically objective is the attraction of a work of art which *is* such a first-person-insight. Like a living person, a young dog, a toddler, or a Northern Californian coastal meadow teeming with life, it draws us nearer without us being able to explain why we feel connected. A striking example for this flavour of poetic argument gives J. M. Coetzee (1999:33) in his reply to Thomas Nagel's (1974) essay "What it is like to be a bat?" by treating a philophical problem through *enacting* a story. In one of the famous scenes of the book the aging heroine argues: "To be a living bat is to be full of being. Bat-being in the first case, human-being in the second, maybe; but those are secondary considerations. To be full of being is to live as a body-soul. One name for the experience of full being is *joy*."

Poetic objectivity deals with the embodiment of existential sense and meaning in its many non-rational guises. These may be pictorial, gestural or palpable in other ways, such as poems, sculptures and music. Feeling in the sense discussed in the last chapters – as subjective experience of meaning and concern through the mutual transformation of one into the other – is not only a category that is universal among all species, but is also a strong, even defining aspect of poetic experience.

We could therefore say that the poetic gesture is the natural expression of the experiences of a poetic-embodied existence. A great work of art seizes us emotionally and by this shows something profound about aliveness. This emotional understanding is also a kind of shared existential experience. It speaks to us individually precisely by its poetic objectivity. But the same experiences of being seen and understood and being able to see and understand are evoked by nature itself as countless naturalists, artists, musicians and ordinary people can attest. Natural beings themselves are poetic gestures that are about aliveness.

Poetic objectivity is weak. This weakness is its strongness. This weekness is the fact that poetic objectivity – the insight into the inwardness of being alive – cannot be transmitted as if it were a thing, but must be enacted. Poetic objectivity is nothing one agent can "have". It arises through the interactions of incompatible individuals forming the process of an impossible whole. We cannot "prove" it with quantification or controlled, reproducible experiments. We can only try to bring it to the observer and let it do its work.

On the other hand, poetic objectivity is stronger than any scientific reasoning because we can feel it and because it can transform our actions even before our conscious minds can recognize it. It relates to our own aliveness and hence we can know about its truth without having understood anything on a conceptual level, and without the representation of an object. The philosopher Ivy Campbell-Fisher (1950) observed: "If I could be as sad as some passages in Mozart, my glory would be greater as it is . . . My grasp of the essence of sadness comes not from moments I have been sad, but from moments when I have seen sadness before me released from entaglements with contingency . . . in the works of our great artists . . . " Poetic objectivity provides something that we might call an embodied-empirical proof. It is weak, but it is contagious.

The Creole of the Living Body

The idea of understanding reality through the lens of a "poetics of relation" was first formulated by the French-Carribean poet and philosopher Edouard Glissant. For Glissant, the world in its depth is not static, is not made from substances, but is always relational. Therefore it can only be understood poetically. "Truth in a system of relations can only be poetical", so he states. This is because truth – or sensemaking, or worldmaking, or cognition – in a system of relations is always one position seen and transformed through the other. It is enactive. It is imaginary. Poetic understanding is someone's (or something's) truth seen as the truth of someone (or something) else. It means to create objectivity, the objectivity of lived-life-in-connection, through imaginary inventiveness, or embodied speculation.

Glissant has developed this view as a theory of culture. But it is equally valid for ecology, and the poetic examples which Glissant happens to chose are predominantly ecological. They are poetical metaphors of living connections. Glissant (1996) has a name for the poetical process of transformative relatons: He calls his poetics a "creolization of thinking". He choses this term based on his observations of the history of a creole culture in his native island of Martinique.

There, the influences of the colonists interpenetrated with those of the geographical place and the culture of African slaves which had been deported to the island. Glissant contrasts this with culture totally enclosed and smothered by colonization through miscegenation ("métissage"). Creolization for Glissant is a mutual transformation and fertilization of self and other without clear hierarchies. Everyone involved has equal standing. All involved are actors and objects alike. They belong to themselves. And they can be means – even food – to everyone else (see Glissant 1997). For Glissant (1996:17) creolization is a striking example of a poetics through diversity.

Creolization is interpenetration. Mutual transformation. Reciprocal Enaction. The invention of novelty through the mutual incorporation of age old constraints. Perception through being open and vulnerable. It is a permanent oscillation in tune with the sensitive bodies of this world which can only exist through a continued sensual encounter including making love and being eaten. In the essayistic poem "Like a Countless Bird," Glissant (2005) suggests a new poetic epistemology that "attunes to the odyssey of the world... it is possible to approach this diverse chaos and to grow by the unforeseeable occasions it contains... to pulsate with the pulsation of the world which finally is to be discovered."

Glissant argues that we have to think in creative paradoxes that embrace their own opposites. This resonates quite strongly with the ecological poetics proposed in this essay: We can only embrace the paradoxes of lived existence if we allow ourselves to think in an embodied fashion, as consciousness in physical form. This is the language of a first-person-science, and Glissant tries to apply this in the poem itself, when he suggests: "Imagine a flight of birds above a lake in Africa, in North or South America..." Creative paradoxes are able to embrace the entire world through the loyal imagination of the possibilities of one singular place.

Relating to the whole while insisting on individuality and locality inevitably shakes one's own foundations. The attitude of risking oneself through insisting that relating does not mean to obey set standards but to meet halfway where both (or all) participants are destroyed and reborn does not attempt to circumvent despair and pain. Accordingly, Glissant calls his poetics the "thinking of tremor". The thinking of tremor is poetic objectivity in action. It is full of the true tremor of insecurity of being totally open and vulnerable. And it is inherently productive and through this continuously reinvents the terms of consolation.

Glissant's poetics is an illustration how inviting contradictions to exist and even to flourish can intensify our view of the world and the relations of all of its parts in a nonhierarchical way. It celebrates the richness of an existence which does not define itself by identities, but by relationships. We must not fight these contradictions or flatten them out. They are the material for life's creativity and the raw stuff upon which improvisation draws. Accordingly, Glissant claims that any personal, literary, and political narative cannot be about "my race" vs. the others, or "culture" vs. natural resources, but rather must be "my particular biography" that relates to a particular place that is a particular habitat for particular species – yet which nonetheless has universal resonances which are felt.

Glissant's thinking shows how the natural history of "freedom-in-incompatibility" can be integrated into a poetics of the world, and how this poetics lends itself to a political view of things. In the center of this stands the certainty that all lived reality, be it physiological, ecological, emotional, sociological, political, economical or artistic, is paradox. Glissant (1996) therefore strongly argues for a "poetics of diversity". We have to accept the absoluteness of the total and the individual at the same time; we have to see that identities are existential but only brought upon momentarily, through the interbeing of relations.

Objective Affection

Glissant's concept of a "creolisation of thought," which so much relies on the admission of contradictions, is also an "ecologisation of thought." Ecology can be understood as the description of a relational whole composed of individuals who thrive on incompatibilities. Living reality is established through the unforeseeable actions of individuals, who are independent agents, but also parts of a whole. Glissant's "thinking of tremor" therefore is also the "thinking of life." It is the "thinking-action-of-the-embodied-living in relation with the other." Productivity can be achieved, mutual fertilization can be done. Mutual acceptance is possible. But no final redemption, no salvation from being embodied. Ecological systems – with humankind in their midst – are sliding from catastrophe to catastrophe as part of their normal process of transformation and self-creation.

A creolization of thinking requires "peership" between empirical reality and feeling. All processes take place inside and outside of an organism simultaneously; they are always conceptual and spiritual, and they are symbolical as well as also always real in space and time. Creative action is the experience of what is alive, as experienced from the inside, subjectively. One might call it "affective objectivity" – a universal and real phenomenon, but one that is also evanescent and resistant to measurement.

Indian geographer Neera Singh has shown the extent to which this emotive power encourages individuals to act and how much it provides subjective rewards for their action. Singh surveyed the material but also emotional relations villagers in Indian rural communities have to local forests they use. She demonstrates that villagers who still cling to traditional commons systems in order to use and protect forests, engage with them in a sort of creolization process. Both become parts of a larger whole (the commons), which is neither human nor natural, but carries traits of both. The relationship villagers experience is material, but at the same time also spiritual. Therefore, the villagers engaged in this creolization which is the commoning process, not only make resources more productive through their commoning with forests. They also satisfy emotional needs and "transform their individual and collective subjectivities" (Singh 2013).

Both, humans and trees are engaging in an active poetics of relating, in which the human affect and the "material world" enter into a communion with each other and alter one another. How deeply natural commons are an example of the creolization process of aliveness I have expressed in Weber (2015), but the depth of it remains to be explored further (See also Tefera 2015 for an example).

It bears emphasizing that "collective subjectivity" extends beyond the human community to include the subjectivities of the living environment – the trees, the supportive vegetation, the birds, the flows of water, the "real" ecosystem elements that human subjectivities actually alter. It is telling that cultures for whom participation in natural processes amounts to emotional engagement in a poetic reality, do not make the distinctions between "animate" and "inanimate" or "nature" and "culture" – dualities that are taken for granted in Western thinking. The basic affective experience of being in a lively exchange with the world, taking from it and contributing to it, is denied by the West's worldview and language: a perniciously subliminal type of enclosure.

Singh calls the psychological-emotional engagement arising from caring for a commons "emotional work." Emotional work includes the material work. It is work inside of the creole space of mutually fertilizing relations. In the absence of this affective dimension, both subject and object lose their paired identities: those working on the land, say, as well as the object of such work, the animate whole. Geographers and philosophers are increasingly beginning to comprehend land and people as a lived reality – a factor of real interactions and an existential, poetic enactment. Also here, only intersubjectivity can sustain healthy subjectivity.

Culture as Enaction of Poetic Objectivity

If we follow the idea of creolization we see that it can equally describe ecology and culture. In this sense the concept is performative: It yields its own creolization between a practice of the body and a theory of culture. This is really important, because it gives us the possibility to close the gap between the two poles of dualism. If reality is creolization, if it is reciprocal specification (as Gary Snyder (1990) says, the "ordering of impermanence"), culture is not a particular human capability, but rather our species-specific way of imagining our existential situation. It is the human way of thinking freedom-in-necessity. In this, it is always intimately connected to body, as it is the paradigm of the paradox as Jonas described it.

This can teach us a lot of what culture can be in order to keep in touch with living reality. In other words, a healthy culture is a co-creative interpretation of embodied existence in all its irrepressible aliveness. That is why subjectivity, cooperation, negotiation and irreconcilable otherness must not be seen as patterns that only humans lay upon the world, as is currently claimed by most cultural and also by the dominating economical approaches. Rather it is the other way round: Subjectivity, meaning-creation, "weak" non-causal interaction, code and interpretation are deep features of living nature. Its most basic principle comes down to the paradoxical self-realization of an individual through the whole, which at the same time is "the other" that needs to be fenced off.

Culture therefore is not structurally different from nature in the sense that it is only human – a feature putting man apart as incommensurate with the remainder of the world. Culture is symbolical transformation, as is also body. Nature, on the other hand, is not determining human culture in a reductionist sense. Culture can not be explained (socio-) biologically, as neither can body. The causal-mechanic, efficiency-centered approach as a whole is mistaken. Reality – be it addressed as nature or as culture – is basically meaning-centered, open to creative change and constantly bringing forth agents with subjective experiences. It is always creative in order to mediate the realization of the individual through the whole. In this sense it is wild (Van Horn 2017b).

Any exchange-relationship involves both metabolism and meaning, and in this way generates feeling. Reality is a process of unfolding freedom, tapping inexhaustible creativity and intensifying experiential and expressive depth, coming into being only through mutual participation, the transformation of one into another. In this sense, but not in any superficial, reductionistic pattern, culture is like nature. This idea is reminiscent of what Theodor W. Adorno (2013) claimed, arguing that good art does not copy nature's objects, but rather follows nature's deep process of creative unfolding, freedom and "non-identity."

Living reality is self-contradictory – and every culture managing to enliven this reality must be contradictory to some extent too. A grazing commons in some remote highland is an ecological and economic paradox, because only by strictly forbidding to use the pasture for certain times, can this resource be preserved and available in the future. From this viewpoint, the inner ecology of the cell and the social ecology of humans seem to be mere levels in a continuous interplay of freedom and necessities. The living world is self-contradictory because it is "a world where all human beings and animals and landscapes and cultures and spiritualities illuminate each other. But illumination is not dilution" (Glissant 2002). Mutual illumination means to shine a light into the poetic dimension.

Chapter 11
Aliveness

> *"There must be a kind of painting totally free of the dependence on the figure–or object–which like music, illustrates nothing, tells no story, and launches no myth. Such painting would simply evoke the incommensurable kingdoms of the spirit, where dream becomes thought, where line becomes existence".*
>
> (Michel Seuphor (in Lispector 2014:xvi))

Culture is not structurally different from ecological transformation processes, but echoes them in human species-specific creative forms. It expresses our own poetic interpretation of the ever-recurring theme of coping with the irresolvable paradox of autonomy and wholeness. That is why human culture cannot control and engineer nature as a passive, non-living object. Human culture and that what we have been calling "nature" are two sides of one thing which we could more aptly call "aliveness".

Aliveness is the process of creating differences which then interpenetrate, of producing separation which yearns to re-unite, and unity which strives for differentiation. And all this happens as the breathing of sensitive flesh, it happens inside ourselves and it happens around us, and is one. Aliveness needs mutuality. It is identity through mutuality, and mutuality through the common unfolding and reciprocal transformation of identities. What we really mean when we say "nature" therefore is this: mutuality in aliveness. This indeed is precious, as life can only be in mutuality. What we find in other organisms is aliveness: ours, and theirs, and that which is the source of all.

Because we humans are implicated in the creative aliveness of nature, our culture must also honor our own aliveness as the best way to foster our own freedom and long-term survival. We must shape our paradoxical autonomy-in-relation according to the needs of a larger whole that is necessary to all life. Poetic objectivity thus claims that we must be able to submit any practice to the question: Does this bring more aliveness into the world? Does it enliven you? Me? Is it a poetic accomplishment? Is it gracious? Does it enhance the complexity of relationships? Does it give depth a space? Does it bring more life? Does it convey an experience of aliveness? Does it make life fuller? Or, with Glissant, does it enhance creolization? And, with Varela, does it enhance the imaginary dimension? With Jonas, is it free in necessity?

A. Weber, *Biopoetics*, Biosemiotics 14, DOI 10.1007/978-94-024-0832-4_11

These are obviously not the questions utilitarians ask. Concerns with value from an enlivenment aspect take individual experience, freedom, growth and health into account, and recognize that a life-enhancing improvement can be most fully grasped by poetic imagination. It cannot be completely analyzed or exhaustively measured. It can be known best through experience and sharing – in the same way the truth of a poem can only be understood from within the core self of a sentient being that uses language as a means of understanding the self of another being. Poetic objectivity is objectivity from a "shared first-person perspective."

The idea of poetic objectivity which complements "the view from the outside" (objectivity) with the experience from within (subjectivity), calls for a first-person-science to generalize this richer kind of knowledge. This "first-person" does not encompass the human ego perspective alone. It also means to give voice to non-egoistical human feelings – as well as to other "first persons" of experience. First-person in aliveness, as the discussion of the last chapters has shown, always already presupposes the second person, the "thou" of other beings.

"Thinking Like a Mountain"

A first-person science (and practice) takes account of the inner dimensions of foxes and fish, rivers and forests, oceans and shores. To take such a perspective means, as the pioneering eco-philosopher Aldo Leopold (1949) termed it, to "think like a mountain." Through poetic objectivity, the first-person view inevitably becomes a second-person-connection. We see through our vulnerabilities, through our openness for others, through our empathy, which we cannot totally control, because it comes into being through the involuntary reactions of our bodies.

We see through the others, with the others, with the furs, feathers, wings of other beings. We see with their stoicism by which they embrace necessity. We see by their existential bravery to only do what is necessary, not more, but not less. We see with aliveness as such, and with the perimeter of absolute values that follow from it, as Aldo Leopold (1949:97) so patently describes: "Amid the endless mediocrity of the commonplace, a crane marsh holds a paleontological patent of nobility, won in the march of aeons, and revocable only by shotgun."

Art is one means of passing that nobility along. Every mode of contagion with life is about all life, and hence is a variant of poetic objectivity. The artist is striving for the utmost objectivity, because she wants to give back life, because she wants to resonate with what has stirred her, what she only feels and cannot truly name, as there is no name, only a further question, and she wants to make that stir others, too. Nobility here means to surrender to the truth of that what is alive. Then the object of thought or the artistic symbolization being worked on becomes alive on its own accord. Seeing a mountain and resonating with it may yield a poem or a picture, which is not the mountain, but a thought in its vein, the transformation of the moutain through sharing. But it might also turn out in the strife for conservation, in the planting of a new stretch of forest, in more mindfulness toward one's own children or oneself. It can mean to create, but it also can mean just to live.

An example of what thinking like a mountain can mean we can find in the later life of French painter Paul Cézanne, who for nearly 20 years stubbornly went out and painted one single mountain, the Montagne Ste. Victoire to the west of his native town of Aix-en-Provence. In all these years, Cézanne came back with hundreds of different mountains, and nowhere among them the true Montagne Ste. Victoire can be found. It is not in the pictures, but in the process of being seen and transformed, in the creative transformation between the real landscape, the person, and the canvases which became landscapes in their own right. The true Montagne Ste. Victoire is between Cézanne and the solid rock massive, and therefore it also occurs between us, the observers, and the images the painter has left.

For Cézanne painting meant not to show the mountain, but rather to become it, echoing its "moutainness" with that in him that could tune in to the same quality. In Cézanne's example we can see that this idea of "thinking like" is completely different from a naive objectivist view of likeness or of representing something in a work of art. To make a work "like" something or somebody in the poetic space means to pass on the energy of having been drawn into the surge of aliveness of that being or thing. This is what happens when we get in contact with powerful paintings, or poetry, or persons.

A deep limitation of conventional scientific objectivity is its inability to advance the protection of landscapes and species, or social justice, or a fairer economy, because it by definition excludes this first-person perspective of other beings. Poetic objectivity helps us overcome this problem by enabling us to rethink our relationship to earth. It lets us properly recognize our human life as a matter of embodied living-within-the-biosphere, blending materiality and meaning in the same commons-based system.

Using this lens to see, we can begin to re-integrate the material, or third-person aspect of reality with the felt, first-person side that is otherwise "hidden within." Both are equally valid and cannot exist in isolation without distorting our understanding of the full context. This relation is eloquently expressed by Gary Snyder (1992) in a couple of short, koan-like verse: "As the crickets' soft autumn hum/is to us/so are we to the trees/as are they/to the rocks and the hills."

In this respect, any careful poetic connection to a phenomenon of life becomes a scientific observation. A beautiful example of this ecological research in the first-person we can find in the poetic genre of "nature writing" represented by John Muir, Barry Lopez, Gary Snyder, David Abram, Robert MacFarlane, among many others. The idea of poetic objectivity acknowledges that our sentience, our aliveness, is a scientific instrument. Many may object that such an idea stretches the definition of "science" to a breaking point because science has traditionally held to the notion of measurement, reproducibility and falsifiability as key elements of the scientific method. The idea of poetic objectivity makes the case for a broader, more reliable scientific methodology that can acknowledge the inner dimensions of living.

A first-person-science of poetic objectivity should attempt to corroborate those theoretical findings with methods which make the felt existence accessible and that also enable the sharing of these experiences. First-person-science considers feeling, expressiveness and meaning to be another important engine of scientific inquiry. Experiential methods are not the only tools, of course, but together with empirical observation and reasoning they are means to refine and share our experiences. They can become objective with regard to the body, which is the common ground of experience in all organisms.

From the Poetics of Organism to the Meaning of Nature

Nature – its principles of contradictions, yielding meaningful experiences – is also "inside" ourselves. It is not too far-fetched to claim that to experience the symbolic and experiential side of our beings and to integrate them into our personalities, we are dependent on the presence of nature – forests, rivers, oceans, deserts, oysters, song sparrows, and hummingbirds. Nature is a twin that animates our symbolic selves. We gather food for our thoughts and mental concepts from the natural world. We feel ourselves through the embodied core selves of other beings. We transform plants and animals into intellectual symbols according to their real or presumed qualities. The snake, the rose and the tree e.g., are each examples of powerful organic images that speak to our human identity, which is why they recur so often throughout human history in our stories, paintings, myths and other cultural forms.

This process works in a reverse direction as well. Nature embodies what *we* are, too. It is the living – and enlivening – counterpart of our emotions and our mental concepts. Only by being perceived and reflected by other life are we able to understand our own. Only in the eyes of another being can we ourselves become a living being. We need the regard of the most unknown. This necessity of building up our identity is one of the most prominent cultural constants in human beings, from indigenous peoples using animal symbols (e.g., in rock art) to the constant appreciation of nature metaphors in contemporary poetry.

Such practices can release those layers of feeling in ourselves that otherwise remain locked up. We need the experience of engaging with a "living inside" that stands in front of us, displaying itself as a fragile, mortal body. We need other organisms because they are in a very real sense what we are in biological and in psychological respect, but they give us access to those hidden parts of ourselves that we cannot see, because we cannot observe ourselves while observing. There is always a blind spot central to the establishment of our own identity. Seen from this point of view, other beings are the blind spot of our self-understanding.

Poetic objectivity integrates this shared equality into the whole biosphere of the more-than-human world. This is the last of the striking paradoxes we shall embrace. To become fully human we need the relationship to that which is emphatically non-human: the interbeing with other living beings. We have to become animal to truly feel what it means to be human, we have to become plant, we have to become stone.

In the republic of innumerable species and existential relational processes, all contradictions are embedded without being flattened out. We could even say that through the beauty, through the searing emotion that natural settings are able to provoke in their human participants, we feel the balanced existence of all those complementarities: That life is a gift and a burden; that necessity must be obeyed to be free; that death is unavoidable. All this is written nowhere, but enacted through the unknowing wisdom of commoning among myriad feeling bodies, plants and organisms.

Plants and animals are not just abstract models for relations. They are the relations in their very enactment. They are the enactment of the paradoxes entwined in these relations. They are closed unto themselves, as any living being is, and at the same time they are open and touchable. Something rests in the middle of their being that is accessible and yet absolutely unfathomable. It is not alien, but it is without limits. They are what Goethe referred to as "Urphänomen".

Human Culture Is Nature

Human semiosis is linked to a general tendency of the biosphere to become semiotically transparent. It is linked to the poetic character of life, and hence is situated at the heart of living nature. This is crucial for our self-understanding, for our understanding of life and of the role we are assuming in it. Not only our theoretical approach to nature but also the practical steps we take towards identifying with it and preserving it are dependent on the stance we take here.

Humans think in symbols and metaphors. Nature has been man's prime cultural obsession for 1 million years. If we look at the archaic and indigenous cultures of history and the few remaining islands of tribal culture, we must admit that here culture and society are conceived nearly exclusively in natural terms. This finding is what inspired the (wrong, but characteristic) term "animism": archaic descriptions of cosmos teem with life. But also throughout western culture, nature forms a deep undercurrent, manifesting in literary symbols, pictorial styles, and architectural principles. Simon Schama (1996: 27) observes that the "cultural habits of mankind have always allowed for sacred nature... All our landscapes and landscape experiences... in reality carry the stamp of our obstinate and inexorable obsessions."

The critical stream in the humanities has always been struggling to engage with these deep obsessions. The evolutionary approach in biology on the other hand has left no room for a poetic view in its own right, regarding the experience of being emotionally close to nature rather as an epiphenomenon, a purely functional behaviour that favours enhanced offspring frequency. The biophilia hypothesis about our evolutionary-based intrinsic need for wild nature which Wilson (1984) had developed is an interesting exception to the general meaning-blindness of both science and the humanities. What Wilson really meant by biophilia, however, might only become fully plausible if re-read in the light of a poetic biological position.

The view of biological poetics I propose and develop in this chapter is able to connect those deep human experiences and their related cultural contexts with a scientific understanding of life. Within nature, those values and meanings that the life process produces, manifest as vital forms. They are observable by the senses. In the bodies of other beings experiences such as abundance and threat, prosperity and hunger, death and birth are not hidden but visible. They are manifest in the appearance of other organisms. Only in the light of a general poetics of the living we are able to set down a position which is able to include the cultural and the organic approach.

In order to fully experience the symbolic and experiential side of our beings and to integrate it into our personalities we are dependent on the presence of nature. Nature does not act as a mirror, but as a transformative imagination of ourselves which gives us our true identity in the same way the mother's friendly gaze sees her child's pain, accepts it, and embeds it into a broader mesh of life-giving relatedness in which is can dissolve. We have seen that 'interiority', hence mind, can be considered as the meaning of the existential embodied realities. In a parallel movement, the embodied organic realities outside, are symbolic for the human interior perspective. Only by the experience of the living depth of nature the own inner universe the poet Rilke is speaking of as 'intensified sky' becomes accessible.

For there is an inner relationship of space that only makes possible the real spaces of the world. This is the space of the inner experience, a value- or meaning-space that measures and expresses real outside spaces in its terms, and that lays a metaphorical and synaesthetic basic grid on the world, with its meshs determined by the existential values of organic meaning. This primordial space is *not* the body. It is the space of organic existence, the absolute space of existential meaning. This space is real not as euclidian topography, but as the synaesthetic and ecstatic space of aliveness.

Nature acts as the main reservoir of organic symbols. These symbols are essential for grasping our own interior in a concrete, embodied form. Therefore, the presence of natural forms is necessary for us to fully understand ourselves. This is the reason for the deep human experience of natural beauty. That such a deep experience has been existing throughout the ages is echoed by many recent works in cultural history that show our species' strong obsession with forests, trees, lakes, animals, and any other natural settings acquiring symbolical powers (Schama 1996; Böhme and Böhme 1996). Oelschlager (1991) has observed that the people who dedicate themselves to preserving wilderness are trying to protect this symbolic cosmos, and not a particular landscape.

Nature embodies what we are, too. It presents the outside of our emotions and our mental concepts. Other beings are partners in our core selves, which we need to become subjects through the contact with others. Other beings undergo the same drama of aliveness as we do. Trees can tell us about what it is to live, because they display the existential spectacle of subjectivity-in-connection in a living form. After their symbolic death in winter they burst into green again, they grow, bloom and bear fruit, without any intervention from our side. Productivity, innovation, harmony, but also decay and failure happen not only to us and our projects but to all of nature. The elements' powers, the becoming and vanishing of beings, the alternation of light and dark embody the processes that occur in our inner landscape.

Finding Ourselves in the Others

In the bodies of other beings we see how the same powers that we identify from our own embodied experience become form. We only know these powers from the "inside". We cannot understand them in an objectified way except in the bodies of other beings. In a Californian Poppy's orange splendour the happiness of all beginning, of unprecedented unfolding, untouched and bright as untrodden snow, becomes real. By seeing and touching the flower, which is another organic being, we can grasp these subjective-emotive aspects in an objective fashion. Nature is the looking glass in which we can see expressed what we are ourselves. In a living tree, for example, we can experience forces which we "know" from ourselves – not in conceptual terms, but as "vital import" (Langer 1953). In the knotty oak, in the flowing sea of grass, lie as many layers of experience as in myself feeling my strength or my melancholy.

I am able to understand my own beginning, my own hope, my own fresh starting as a facet of more general instances of life. I partake in a general condition of life. I have called this the "conditio vitae" (Weber 2003) – in contrast to the term "conditio humana" which has been coined exclusively for the human situation. In the human condition everybody is confronted with a situation he is "thrown" into and cannot understand. In the "conditio vitae" the organic reality links us to every being and to every other living being's experience and expression. We are brothers and sisters to everything which experiences poetically and which hence is alive. We share, on an empirically biological level; we are part of a greater connectedness, of a living cosmos, subject to a general principle of life which carries us over episodes of individual loss. We are no longer "thrown into" isolation, as in the human condition, but also "supported" by more than human life.

When we accept that our own subjectivity-as-becoming is always a primordial relation, that subjectivity itself is a dialectics differentiating with a very ephemeral hand the ever-changing demarcation line and trading schemata between self and other, then we find in animate nature not just a preverbal wealth of symbols of our own experience. We can also see in them the conditions to be a subject. Living nature, the totality of other beings, is the paradigm of the relation of the-self-in-the-other, and thus our experience of them might be the archetype of relatedness as such.

To make poetic objectivity possible, we need the existence of other beings. We must not colonize them, if we do not want to silence the powers of our souls. The distinctness of many of our experiential categories might only be possible because in wild nature, in *natura naturans*, there is this form of subjectivity untouched by man, which has brought forth us, and which still guides us as to how to confront our own embodied existence. Here lies the path by which dualism can be healed.

By being transparent in itself, living nature reveals the functioning of the cosmos that has brought forth life as the paradigm of the self-bearing, self-producing, flowering plenitude of a fragile and precarious exchange that is always doomed to fail, but that at the same time will always be replaced by new growth. This is the only adequate model of the soul.

The deep cleft which has opened up between us and other beings, between the world as we experience it and the world as we describe it, closes again, and for the first time for a long period we are welcome. Plato had suggested that for every term, be it as abstract as can be, there was an *eidos*, an archetype in the empire of ideas. Certainly, Plato was not completely clear at this point. The empire of ideas does not lie beyond, in an ideal world, but is anchored here, in the bodies of plants and animals, in the buzz of the bees and the shape of the circling raven.

Chapter 12
Poetic Space

La poésie est le désir demeuré désir.

(René Char (in Midal 2009))

We radically need to think ourselves – our bodies as living beings, and any other living beings, too, our intersubjective relations, the process of our identity – as self through-other, self-as-non-self. Through this, the poetic viewpoint enters into any view of objectivity. A poetic relation is the only possible relation to understand other through self. In a poetic relation there is familiarity and strangeness and only both together make sense. Poetic objectivity is the objectivity learnt by dwelling together in poetic space.

Reality is this poetic space. It is not in truth "only" matter, nor is it "only" mind, or discourse, nor is it a dappled assemblage of both. It is rather both at the same time, and through one another. It is both, making one side accessible through the other, and only through the other. It is both as the core self of every individual organism is neither a biochemical process nor a detached concept alone, but a concretisation of existential concern through matter, and the unfolding of inwardness which matter hides.

The world in its entirety is this poetic space. But we tend to forget this. Rational thinking has omitted that there is a rationality of the living flesh, which is not chaotic, but follows the laws of biological growth, which very well knows that every being is a subject only through being through others, giving life and transformation, and needing the same in return.

Other beings do not know this in the manner a philosopher believes he knows, in the form of knowledge *about* the world. The song-sparrow, who is chanting his melody on the shaking stalk of a late march Cliff Spurge blossom in a ravine leading down to a beach in Northern California, his song blown away and brought nearer again by the gusty wind, does not know. He does not sing about reality. He sings as reality, as a force expressive of the imaginary dimension which is our true dwelling space, which comprises bodies and ideas and which provides the energy of incubating one with another.

Everything happens within this poetic space, every encounter, every felt meaning, every expression of an existential concern, every connection, every metabolical act, every transformation like those our liver is concerned with in every second. But this

A. Weber, *Biopoetics*, Biosemiotics 14, DOI 10.1007/978-94-024-0832-4_12

meaning space neither is the one "true" reality which then, as a sphere of meaning, would prove that the world is "ultimately mind". The poetic space is made from meaning, but is still matter. It is always inside of us, but in order to connect to it, we use our senses, our skin, our eyes, our mouth. It is material through and through, but it is neither "mere matter". It is the flourishing matter of biological desire that in every instance yearns for unfolding, for touch and connection, and bitterly hurts as these desires again and again necessarily must remain unfullfiled. As we humans know, we are body and inwardness at the same time, and we are even not only one body, but several, a multitude of billions of symbionts such as the many species of bacteria within ourselves.

The world is a physical resource and a three-dimensional space, but also an emotional reality – an "inside" as well as an "outside." Individuals and the biosphere encompass both material processes and meaning relations. Together they constitute lived experience, which from the inside of organisms is subjectively "felt" and from their outside exhibits itself as "sensuous" and "expressive." This poetic space is not to be confused with "spirit" (inside) and "body" (outside), but rather both conjoined as metamorphic material that is always meaningful.

This idea breaks with any notion of primacy of either matter or symbolic relationships, and so in this radical way is non-dualistic. There is no outside to poetic space because the poetic space encompasses both organic and non-organic matter. At the same time the imaginary scene of this poetic space can be subject to transformation from both "sides": through material manipulation but also through imaginative creation. The poetic space is open to new interpretations, new utterances of self-expression and can be transformed in such a way that real change in the world takes place.

We can escape the habit of thinking in identities if we accept that everything is in continuous change – as the body that exchanges all its atoms with the environment every few years through the process of metabolism. Any process goes through "good" and "bad" states. Process is not stable, but rather a constant fluctuation. So history has no clear direction towards the "better," although this is taught in monotheistic religions and practically attempted in neoliberal economics. Rather, we can see that the only quality that really grows over time is the amount of different experiences – felt depth – in biosphere over time.

Life longs for more and more experiences about itself. And also we do. We desire to become alive more deeply. We have an inborn instinct for this aliveness, which tells us how to make life – life as such, not only our own individuality – deeper. It is a drive that tells us to do the same as the world does: to intensify our experiences, to extend our knowledge of ourselves and others, to unfold new capacities, to strengthen bonds, and so on. One might say that this process is about learning to respect and learning to love.

The poetic space is the meaning which is not there, but for which we are yearning, and in this yearning it becomes present as an abesence. It is the meaning which calls us to unfold it through our actions and perceptions.

Held in Reach by the Same Star

At this point of the argument let me come back once more to Bateson's use of the concept of abductive logics. As you remember, for Bateson the most apt logics in the realm of other living beings was reasoning through the "pattern that connects": Grass is mortal, humans are mortal, hence we are grass. By saying this, Bateson states that we dwell in the same poetic space. Grass and humans are agents and forces in this space, in which they can mutate into one another. They are not really different, as all which happens within this space is valid for all living beings. It is the most basic truth about what it is like to be in the world.

In using "the pattern that connects" as a name for this, Bateson did not only address the sphere of matter, but also that of meaning. The pattern that connects is what relates us to the poetic space and which makes us being a part of it. It is the symmetry of a crab's body washed on the pacific shore between old boulders wet with algae, which connects it to the embodied regularities of other beings, and to our own bilateral symmetry. At the same time the crab is, through this symmetry, a statement about the necessary fate of all lifeforms, which need to reach out and look for food, for shelter, for others. It is also, as a piece of mineral interspersed with organic patterns, a transformation of the earth's crust through the desire to be a self.

All this is one, all this cannot be seperated into a technical section (mandatory) and a poetic curriculum (voluntary). It is always both and one through the other. The minerals in the crab's shell will be ground to fine dust by the waves and later will settle as marine sediment, becoming rock again. But they have been the vehicle of desire and contact, of exuberance and of tenderness.

Understanding the other means to see her as part of the poetic space. Seeing someone as part of the poetic space means to see someone as ourselves. We have seen in the last chapters that through the mirror systems and related neuronal and behavioral affordances living beings automatically meet one another in poetic space. They cannot help it. Thou is part of the poetic space, as is that I, which can only be discovered through the other.

We are never alone in poetic space. And exactly this is the reason why De Jaegher (2015) could observe that "participatory sensemaking allows us to reach into each other's self-affection." Not only humans do this to other humans. The grand role of nature – the total of beings, other and same – is that it becomes part of our own self affection. We feel ourselves in and through the others. Who these others are, what they are, and also *if* they are, is by no means trivial.

In her poem "The Silence of Plants" the polish poet and Nobel laureate Wisława Szymborska explores how we are meeting these other beings in poetic space. Szymborska's poem seems like the long version of Bateson's epigrammatic "Men are Grass", which already is a very short poem. Szymborska advances further into poetic space, led by the fact that this is not a storage of simple truths, but a labyrinth without ending, in which we get lost and through getting lost become ourselves.

We end up with a faint understanding, but it is understanding only insofar as it is yearning for something we do not really understand. Szymborska's poem is so instructive for this that I might be excused for quoting it here in full.

The Silence of Plants

A one-sided relationship is developing quite well between you and me.
I know what a leaf, petal, kernel, cone, and stem are,
and I know what happens to you in April and December.

Though my curiosity is unrequited,
I gladly stop for some of you,
and for others I crane my neck.

I have names for you:
maple, burdock, liverwort,
heather, juniper, mistletoe, and forget-me-not;
but you have none for me.

After all, we share a common journey.
When traveling together, it's normal to talk,
exchanging remarks, say, about the weather,
or about the stations flashing past.

We wouldn't run out of topics
for so much connects us.
The same star keeps us in reach.
We cast shadows according to the same laws.
Both of us at least try to know something,
each in our own way,
and even in what we don't know
there lies a resemblance.

Just ask and I will explain as best I can:
what it is to see through my eyes,
why my heart beats,
and how come my body is unrooted.

But how does someone answer questions
which have never been posed,
and when, on top of that
the one who would answer
is such an utter nobody to you?

Undergrowth, shrubbery,
meadows, and rushes . . .
everything I say to you is a monologue,
and it is not you who's listening.

A conversation with you is necessary
and impossible,
urgent in a hurried life
and postponed for never.

Poetic space always is opened up by the other beings. And poetic space is what we always have access to. We have access to it because it calls upon us. Indeed, it is the dimension in which we experience, but in which also the bodily processes underlying this experience happen. It is the conflating place of both of these dimensions, but not because they merge, but rather because they emerge from this unifying realm. Because of this, because of the fact, that reality ultimately is about the poetic horizon of relationships in transformation, it is futile to search for an ultimate reductionist explanation of sentience, or experience.

Access to the poetic space is granted to us by the means we already have, which are immensely powerful and extremely limited at the same time: our own inner experience, and the experience of the other's expressiveness and gestural gestalt. An account of ultimate reality, which is aliveness, can only be created from within aliveness. It works its way through expressiveness by means of contagion with aliveness, with "vital import", as Suzanne Langer (1953) had it, and through the "circle of the gift" (Hyde 2007) of passing on the beautiful contagion with aliveness, which in others triggers the existential desire to be more alive.

I don't want to exclude theorizing about the physical conditions of aliveness, but utter a caution against the inherent dangers of objectification. Accounts of aliveness are incomplete and miss its – and our – subject, if they are not also a reverberation of personal experience and transformation, and if they do not to intrinsically include the first-person-perspective.

We can understand the dimension of the core self only by means of the core self. The core self is not a secondary derivative of biophysical processes. The core self is a window to the dimension of the world where matter and mind are not separate. It is a reality before the separation and it is more primary than the separation. In order to access the core self, all we need is to speak its language. This is nothing extraneous, it is a language we speak from the very beginning, as all language starts with the expression of our core selves. It is a language which dissolves if we try to formulate it in mathematical terms. It is a language which communicates only by expressive presence. The core self is accessible in the presence of a flowering St. Anne's lace. Or in a hare's struggle against death. In a child's smile. In the cricket's late autumn song. In the hummingbird, hovering amidst liquid air. The core self is nothing to be extricated from body, it *is* body, in its material presence, but from the inside.

This view by means of modern biological insights, refined through understandings gained by embodied cognitive science and molecular biology, seems to echo the "empirical spirituality" developed by the British romantic poet William Worsworth, for whom "the spirit must always be a palpable, sensible presence, seen, touched, smelled, heard, tasted . . . " and "'beyond' is only reached 'through'" (Roszak 2000:111). The inner world is totally present, but accessible only through the flesh, through embodied experience, through the transformations of perceptions which our – an organism's – senses allow. It is ecstatically present and perfectly veiled.

To access it, we need to be at the same time absolutely naïve and intuitive, but also highly cultivated and artistic. We must "let the soft animal of our body love what it loves" (Oliver 2004) in order to establish the connection. And we must carefully chose the language and culture of that connection. We must imagine afresh which is clearly given, but which is in reach only through being created, felt a second time, imagined, calling forth, as is a poem. If culture, or a particular practice within it, achieves this, it enlarges the poetic space without enclosing it. Such a culture is not different from body anymore.

Nature is Inside

In order to explore the flesh of the world we need to be poets. And we already are. In truth every being is leading a poetical existence by mean of its authentic feeling and genuine expressiveness. We are all born poets in this sense, but many humans unfortunately unlearn it. The true activity of poetic creation is to continuously make new connections within the flesh of the world, letting it grow, swell, and feel itself. Aliveness is ongoing coming-into-being. This can happen by any practice which favours the mutual fertilization of self and other. It is what we immediately experience as life-giving, and what makes persons who have this generosity, so attractive.

Therefore, also linguistic poiesis is more than word magic or suggestion. Pictural and Musical explorations of vital import reach beyond mere symbols. They enter into the core self. The magic in a poem is its force to seize us with its meaning in a premodal way, to speak to – or rather become active part of – our core self. How else could the nonspatial sphere of a poem be able to create physical atmospheres and make the skin stir and sing? It can do so through the way a poem is able to be contagious with life. Everyone who reacts to a poetical expression is brought into a state in which her core self reacts to the aliveness evoked. Poetic expression which achieves this mobilizes something which affects us although it may not be understood. It creates a mode of existence reminiscent of the stirrings of happiness, evoking melancholy, shy hope, and hesitating consolation.

Artistic mimesis strives to have the same vital import as that which it translates has. It transports the freshness and sincerity of the real experience. Such a meaningfulness forecloses any control through discursive rationality. To make the unspeakable speak it has to be transformed into another unspeakable dimension. How should the absolute presence of an individual human's face be understood else than as the embodiment of all gestures that can ever be expected from this person? An accomplished rendering of her face needs to incorporate exactly this particular openness, an individual flavour of infinity.

What really responds to a poem – or any work of art which transports aliveness – is the embodied self, not the personal consciousness. Poetic symbols embody vital import. We can see this in other forms of embodied metaphors, as e.g. happens through the embodiment of interpersonal relationships through positioning figures or real people. Without description or explanation, positioning can elicit strong emotional reactions, emotions which then can be reappropriated and worked upon. In that sense, all artistic expressions are core self work, and therefore meant to work with trauma, because trauma is that which threatens aliveness. They help discover the lost emotions, which are fragments of the core self and have become muted, but still are there and still are asserting themselves in the labyrinth of poetic transformations.

Poetic expression enables us to incorporate these dimensions of reality. It means to see with words, with images, with bodies, which can become sensual organs discovering literal truths. Some souls are icy because in their presence we shiver. Some hearts are warm because in their embrace we thaw. In relation to an existential background both, the real ice and the cold soul, the gentle heart and the kind sun rays, have the same effect. In poetic space they are equivalent. More, they are indistinguishable, as in that space of feeling only vital gestures are real. In poetic space words can kill, and caresses are able to heal – as can, as we have discussed above, kind marital relationships. Already 70 years ago Gaston Bachelard (1990:33) had argued that poetic language works through an existential equivalence within a space ripe with vital metaphors. Bachelard accordingly believes that "the objective attempt to start a fire by friction is based on utterly subjective experiences . . . love is the first scientific hypothesis for the objective reproduction of fire . . . love is nothing but a fire which needs to be passed on. Fire is nothing but love which needs to be captured."

The qualities experienced through poetic language are real, because they effect our organism. They effect us not on a bodily level alone, however, even though the effects might manifest in a bodily way. The qualities are premodal. They are important for the core self which exists beyond the symbol/body gap in poetic space. Becoming manifest in poetic space explains why the seeming paradoxes of poetic language are so elementary: they represent the paradoxical union of self-and-other. They capture the faint break inherent in relatedness as such, that what Szymborska calls the conversation which is urgent but impossible. It is impossible, but that impossibility marks the whole endeavour of poetic expression, not only of communicating core selves, but also of art. Poetic worldmaking means to see with that which is *not* an eye, because it is our vulnerable wholeness, our precarious aliveness. It is listening with one's core self, as the French poet Claude Vigée does, who in a poem "hears the greening of a young walnut tree". To *hear* green a tree is not a hallucination of our senses, but describes our baffled reception of the first tender marks of spring.

An existential symbol *is* what it means in material terms. The menacing gesture of a looming predator is the symbol of impending destruction for the mouse, but it is so because the gesture has a perceivable form in time and space, an element of reality which will lead to the outcome which is feared in a symbolical way. The score of a jazz tune which I listen to on my stereo system (the "black classical music", according to singer and pianist Nina Simone), is a poetic statement through being an assortment of solid-state resonances, ac voltage and air waves. It is material, and it is meaning. It is solid, and it is abstract. It is here and now, and it is totally detachable from time and space, by having become a poetic expression *on* material solidity.

By these means Homer's "wine-dark sea" (Langer 1967) can mutate into the oceanic desert, and inversely the sandy bleakness can become an endless water in which the dunes flow, and both can stand in secret correspondency to the soul. The infinity of the real desert finds its echo in the unfathomable inner being. Poetic space means that there is an inward spatial relationship which comes before the real spaces in this world. Here endlessness happens first. And only by knowing it from the existential inner reality can infinity be discovered in the world. The world and its relations are not different from the possible scenes and constellations which the basic process of self produces and hence knows, as it is already the primal relation of autonomy and whole. The space which knows this is not the body. It is the space of embodied existence, poetic space, the absolute space of existential values.

This poetic space therefore at the same time is the most simple and the most enigmatic of all possible spaces. It is entirely of this world, but it is this world in the absoluteness of meaning. To experience these meanings does not load the world with individual projections, but rather leads to being sensitive to the "inscapes" of things and bodies, to the existential symbolics of the individual and the whole of which we are a part.

Many symbols of nature in this perspective are not metaphors, but crystallized insights into forms of deep connection, ideas which do not refer to a platonic beyond but to an embodied here and now. The philosopher and writer Mircea Eliade (1990:105) observed that the physical sky through its concrete relation to ourselves reveals the ideas of transcendence, power, and eternity. These ideas exist in an absolute manner in the material sky, as it is high, vast, and home to forces that we cannot control. Therefore it *is* the poetic space of that which is unreachable, endless, and powerful. The blue of the sky is primordial, as it connects our experience directly to its meaning. It does not show anything, but it is.

The way in which the poetic space becomes perceivable is the poetic dimension. Let me shortly sum up the definitions of this poetic dimension we have come upon during the course of this book. And let us be reminded that they are only a footnote to an age long aesthetic search in which others have set the scale, as Friedrich Schiller did with his definition of the beautiful as "necessity in freedom" or Friedrich Schelling in his view of it as "finite infinity". To enter poetic space, we need to embrace paradox. We find this in

—the whole in individual transformation
—inside seen through an outside
—a form which conveys a feeling
—the translation of body into subjective meaning
—a thing which physically has the same effect as what it means
—the expression of something through which it is not but with which it is intimately connected
—the whole in an arbitrarily small fragment
—something emphatically confirming itself by not remaining itself
—a simultaneous modification of the individual and of the whole
—an individual flavour of infinity

As we can seen, these definitions of the poetic dimension all can also be used to describe what a living being is and does.

The Ordering of Impermanence

What Merleau-Ponty calls the "flesh", and what I try to describe here as "poetic space" the American poet and ecophilosopher Gary Snyder names "wildness". To Snyder "wild" means to become self through other. Therefore wildness (and the place it self-expresses, wilderness) does not yield formless chaos (or oceanic idyll) as which it is mainly painted through the dualistic mindset (the evil nature which must be conquered or the sweet nature which gives healing). What is wild is the process by which self reveals itself through other, and through which it therefore defines the fundamental laws of ecology *and* of the soul. The wild is the principle of how self is transformed through other, and how individuality becomes symbolic of the whole, allowing it to realize itself in totality through a specific fate.

To Snyder, ecology is governed by the principles through which individuals realize themselves as matters of expression of the whole, and through which these individuals at some point must give themselves back to that whole. The laws of ecology therefore are this "compact" which galvanizes the rules of the flesh into concrete forms, bodies, and experiences, into concrete niches, species, behaviors, and relationships. Snyder (1990:4) cites the "compact" coyote and ground squirrel have with one another in which one needs to assume the role of hunter, the other of the hunted. These are likewise the roles of dependent and resource, or needful and needed. Concerning our own part in the ecosystem, and our own wildness, we best adhere to these principles. And for a human adhering to them means to imagine and recreate them in our own feelings. As Snyder (2001:1) says the "workings of the human mind at its very richest reflect this self-organizing wildness."

The rules of individuation, expression and transformation, which govern the poetic space, manifest as the multitude of species, phenotypes, temperaments and processes, which we scientific ecology describes. Internally, the ecological features reveal a second set of rules which are about poetic objectivity, about the laws governing exchange and experience within poetic space. Both are, on a still deeper level, nothing but the "ordering of impermanence" (Snyder 1990:5), the whole which is its own part. For Snyder (2001:2) "The world is constantly in flux and totally mixed and compounded." Creativity is an agent of seeing afresh by means of what is very old.

Snyder (1990:3) describes these rules as "the etiquette of freedom". With this he grasps the paradoxical entanglement of that which is necessary in order to enable that which is free, the "dependent freedom" Jonas introduced. There is an etiquette, a morals, of participating in poetic space, which is not about rules of conduct, but about principles of self-realisation which are, more fundamentally, principles of other-realisation, principles of generously giving away, which on the double plane which poetic space always assumes means a more generous way of imagining, of making space for the *surplus* that always wants to enter.

All this is wild. Wildness means that there are unchangeable rules governing the possibilities individuals have in partaking in poetic space, and that, even if they try to evade these rules, they partake in them nonetheless. It means that "the world is our consciousness, and it surrounds us" (Snyder 1990:17). It surrounds us, because in our core self we are poetic space, as the living world as such is, a pragmatic going on about our simple lives and a poetic statement about the whole at the same time.

Language is nothing different from ecological processes. It partakes in the multilayered exchange of bodies and meanings, mirroring and transcending it, as bodies also transcend and mirror it, and as any partaking is just that: mirroring and transcending, specifying in reciprocicality, being I through being you. The world is already poetry, if we could only listen, or feel, and poetry is equally part of the world, if we could only listen. If language is wild, and the world is wild, and language is the wildness of the world, to imagine through language, seeing with language, means to perceive with the wildness of the world. This is what poets explicitely try to do. It is also what every living being involunterily always does, as he also expresses his core self through embodied gestures.

We – and all living beings – are always already wholly inside this poetic space. We only cannot always see it, and sometimes we do not see it at all. To realize the poetic self means to realize that it is emphaticaly not about me, and that the I is about the all-encompassing whole. Caring for the poetic space is not achieved by celebrating the poetic dimension of reality but through being wild. Being wild means to accept that we are participating in order and in impermanence at the same time. Caring for the poetic space means obeying the etiquette of freedom, which is not about me, but about other, about keeping this space open for the field of the work to playfully unfold.

Poetic existence thus is a practice of being which helps the "wild" qualities which are inherent in poetic space. It cares for aliveness by granting the other the space to unfold. It remembers that individual aliveness is only possible through the whole, and that – as painful as that is – it is not a particular individuality, but aliveness as such, what reality desires.

In this respect, as part and origin pf poetic space, every organism is a holographic verson of the whole. Every organism, itself already being a multitude of selves, is the whole in an individual expression. Participating in the poetic space is the work of the core self of an individual. Being in the poetic space is being in the core self. It means to be part of the core self of what there is. That is the lesson we can learn through being a feeling body. It is the lesson which we know from the very beginning and which at best we need a whole life to understand. It is the lesson we can only learn when we answer to a question with another, deeper question. If we quench the yearning which is poetic space not with consummation, but with a deeper desire.

Chapter 13
Conatus

Everything expresses to different degrees the power of life that Spinoza calls the life of God.

(Paul Ricoeur (1992):317)

The perspective opened up through Biopoetics equals the shift in modern physics when scientists realized that an observer is entangled with the object he watches. The development of relativity theory and quantum mechanics led to the revolutionary awareness that in physical science there is no such thing as objective reality. But physicists working with large electron colliders and highly sensitive measuring devices ar not the only ones who can realize this entanglement. From an organism's standpoint, this is an everyday's experience. Objectivity in lived life is entanglement.

We do not need complicated instruments in biology. Every being's first person perspective of meaningful concern proves this entanglement all the time. What modern physics has discovered for living beings has always been the center of experience. It is an emotional reality. To acknowledge it, we do not need empirical investigations about the degree to which the biochemistry of organisms produces biophysical states which can be described in particle physics terms as quantum states. We only need to take seriously the entanglement we can feel, which binds us to others in poetic space. The only realistic experience of the quantum world and its entanglement of observer and observed is present in the perspective of existential feeling.

To close in on the theoretical advances in physics has won over biology, therefore means to incorporate the first person perspective of feeling, expression and meaning into biological sciences. A biology, which is willing to face the entanglement of subject and object that quantum physics has shown to be the basic layer of reality, needs to become poetic. Many still refuse this, however. They are trying to keep biology newtonian and to conserve a perspective on reality physics 100 years ago not only has given up but shown to be fundamentally wrong.

In quantum and relativity theory, reality is always entanglement with the matter a researcher tries to describe. Quantum mechanics was developed, when researchers like Niels Bohr and Martin Heisenberg realized that the measurement of a subatomic entity, say the spin of an electron, does not yield an "objective result", but provides an outcome which irreversibly changes the measured object through interfering

A. Weber, *Biopoetics*, Biosemiotics 14, DOI 10.1007/978-94-024-0832-4_13

with it. What is perceived is transformed. Bohnr and his colleagues understood that knowledge can only be gained through taking part, and consequently through transforming the particular area of the world that is meant to be "measured".

In the same way, in relativity theory, Einstein and others demonstrated that there are no such things as objective space and objective time: both are functions of the observer. Her own position is relative to what she describes. In these theories, not longer objective reality is described, but the nature of the entanglement itself becomes the centre of scientific understanding. Physically, subject and object are no longer separate. They mutually bring forth one another. The relationship becomes the focus of enquiry.

This is exactly the situation which Biopoetics describes for organisms. Biological perception – which starts with the felt experience of existential meaning, can only happen through entanglement of self and other. Self and other are created – and recreated – through this experiential entanglement. Biopoetics takes this finding out of the abstract and technical description of physical science and connects it with the lived experience of entanglement. Biological entanglement happens emotionally and experientially through sharing aliveness with and relating existentially to other living subjects. That means: Through being entangled in poetic space we can understand modern physics from the inside.

On the one hand, biological entanglement is metabolical and therefore material. It is knotted to the ingestion of other and its transformation into self, and the transformation of self into other, so elegantly visibly in the Krebs cycle. Through metabolism, an organism reveals itself not to be an island to itself, but rather a rolling crest on the wave of a transformative process, by which the same matter (say a carbon atom) becomes incorporated into the bodies of diverse beings. Feeding means to incorporate the matter of the world into one's own body. Breathing – taking up oxygen and exhaling the body's carbon – means to send one's own body substance on a journey through the substances of others. What is breathed out as CO_2 thus is not the ingested carbon, as the still often-used motor metaphor would imply. It is part of Self.

Towards the "Quantum Leap" in Biology

On a material basis, the living being which is in contact with the world, *is* this world. The segregation of self is only possible through this continuum. At the same time self is also the negation of this continuum. This relationship reminds of the quantum physical continuum between researcher, experiment, and reality, which is determined by the former. The distant microparticle which is to be "measured", and only through this measurement is defined, acts as if it where part of a continuous arrangement of matter, part of a compact which comprises reality as a whole as a grand metabolism.

Organisms act according to the underlying deep entanglement of all components of reality. They make them visible. Subjects have a direct experience of this entanglement. We call this experience feeling. Feeling is the experience of entanglement on two levels. It is the concern of self which needs to close itself against other in order to exist, and which has to partner with it, also in order to exist.

Through this metabolical emmeshment, through the process of mutual transformations, biological entanglement inevitably gains a perspective of metaphorical imagination and hence poetic creation. Poetic creation can be described as the meaning dimension of what happens on a material level. Continuous imagination is the inner complement to biochemical processes. Entanglement therefore happens not only metabolically-empirically, but also through experienced meaning. The fact that living beings experience existence as concern, and as existential meaning in expressive terms, as joy of unfolding and as pain of dissolving, is our direct experience of entanglement. It is the ground zero of subjectivity which already a newborn experiences, and every living being does as well. We know what it is even before we know anything else.

When physicists proved the entanglement of the research subject with the objects they measured 100 years ago, they could not make sense of this finding. Their confusion was not only due to the fact that what the researchers observed existed completely beyond the physical explanation framework, which then was still newtonian. The results of quantum physics and relativity were baffling also because they contradicted the understanding of organism which had been achieved until then, which was also newtonian. Biology was obsessed with cause and effect, explaining beings as tiny machines. Different to physics, it somewhat still is. The findings of new physics therefore are more problematic in biology than they are in physics. This shows how deeply our image of organism is not guided by what we experience as an organism, but by what we think an organism must be.

In order to understand physical entanglement, we need to come to terms with the fact that we have a primary knowledge of it through embodied feeling experience. Physics needs biology here, not the newtonian biology of genetic machines and deterministic selection rules, however, but a new account of biology as first-person-experience of being alive. Advanced physics needs the biopoetic viewpoint to understand what it has discovered: the mathematical aspect of what living beings experience as the feeling of being in connection, which is the feeling of being alive.

Seeing with Metaphors

Organisms see with metaphors. They perceive with metaphors. They can only see and perceive and orient in the world because these metaphors allow for relating to what there is. They continously enlarge the space of what is possible. From this viewpoint, breath is not only the processing of oxygen through the metabolic network to "burn carbohydrates" (it is also this, no doubt), but a fantasy about the possibilities of an individual's own existence: It yearns to incorporate that, from what it consists, to liberate some other part of itself, and hence imagines its own existence as the atmosphere and its carbon content. Through the metabolisms of individuals, the air imagines itself as pulsating body.

We need to do more than search for quantum states functioning in the smallest dimensions of organisms. We will not find there what we are looking for, because what we are looking for is always here, right in front of our eyes, and behind them. We only have blinded ourselves to its most obvious presence, as we have banned felt experience from science. For some it seems promising to explain life through intederminacy by understanding the functioning of certain parts of organisms (particular in the brain) as "quantum" (Hameroff and Penrose 2014). More mechanistic thinkers challenge these attempts as are hard to prove and vastly hypothetical. They might be spot on the right track. But both sides need to incorporate the perspective of inwardness. It does not make things easier to explain them as "quantum", because this is still an enigma in physics. It is doubtful if we should explain one conundrum with another. (Richard Feynman said that anyone who claims to understand quantum physics has not). If we can describe a biological process in quantum mechanical terms, we have still understood nothing of the experience of inwardness and of the existential concern that comes with it.

Imposing quantum think on biology has the danger of making the problem bigger than smaller, and to shift it to another area of reductive science instead of adding a totally new level of inquiry. Physics itself is still struggling to understand what it really means to live in a world where there is truly no linear space and no chronological time. It does not serve a lot if we push this enigma further down in explaining parts of biology as being quantum, as the failure of Eccles' (1994) model of consciousness shows.

In all those attempts to extend the boundaries of causal explanation (and hence to avoid stepping into the consequences engendered by final explanation, namely the opening up of a sphere of meaning), the enigma remains. This enigma is the problem of how material "substrate" and experience are related, the famous question posed by Kant of how "external knowledge is possible" for a cognizing subject, or, put in a slightly more modern form, the "hard problem" of neuroscience, as David Chalmers (1996) has termed it. It is the question how matter can have inwardness. It is the enigma of how we can understand that we are matter with inwardness.

The "quantum leap" for biology therefore does not consist in the implementation of quantum measurements and calculations in a future theory of consciousness. It consists in stepping back and acknowledging that what we do is already fully entangled with everything else we perceive, and that this entanglement does not happen on a computational level, but as feeling. The semiotic nature of life, and its poetic creation of new imaginative interpenetrations of body and sense need not to be complicated by being pressed in the reductionist framework, even though this be a wildly enigmatic application of quantum mechanics. Rather it is the other way round: Biological experience, which we are having first hand, in its particularity may be able to confirm the wild findings of advanced physics. What makes it able to be "quantum" is its semiotic nature – the fact that in order to exist a body must self-construct and feel as a standpoint of concern, which means to be a focus of the whole universe (the total material continuum) in an individualized standpoint, which is self-creating according to meaning.

We *are* connected to reality, and we *can* have knowledge about it – although this knowledge is not absolute, or empirical in the old fashioned sense of absolute truth and control. But it is also not veiled, or impossible, or utterly fractured, as has been the thinking fashion for the major part of the twentieth and the beginning of the twenty-first century. We know by taking part, and by taking part we inevitably alter anything we know, and we can only know it by altering it (as the incoming light needs to alter the conformation of the rhopdopsin stacks in the retina in order to be made visible). We can know reality as the quantum physicist can know a particle – by invariably interfering with its course. But this does not mean that we are not in contact with the truth. We are in contact by distorting it – as any contact is distortion, but as distortion is also creation, and the enabling of new contact.

Physical Body as Absolute Space

To understand how our own existence partakes in quantum reality means to understand how organisms are generating feeling through their continuous self-creation by transforming that which they are not into themselves and by alienating that which is self into that what is other. It means to understand that every organism opens a poetic space by actively tangling with others.

As any bodily process has an existential meaning, every individual always is part of this poetic space, and is continuously (and potentially endlessly) enlarging and interpreting it. In addition to searching for specific and state-of-the-art quantum physical explanatory frameworks which are supposed to help us explain phenomena of meaning as e.g. mental qualia, and consciousness, we should implement a broader first-person-research agenda to complement these finding with accounts from inside. Varela (1996) and in his wake Bitbol (2010) have suggested a naturalized phenomenology and first-person-ecology approach. For this task, Varela has supplemented conventional mapping of brain activities through electron spin resonance by individual accounts of what the research subjects have been feeling.

We need not invent bleeding-edge procedures in order to find the "missing link" between matter and inwardness. Through poetic production in rituals and the arts *Homo sapiens* has since a long time developed ways to express what existing means from the inside of being. Artists have often described a desire to capture this as the moving force of their work. They try to capture feeling, the glowing core of being alive, by creating more of that which is alive. The american writer Henry Miller (1981) thought that "art is not about anything but the meaning of life". The german painter and arts philosopher Paul Klee has expressed the same in a different manner when he said: “Art is a simile for creation. Each work is a parable, as earthly things are cosmic parables.”[1] (Klee in Schuster 2008, see also Midal 2007).

Art is an exersize in shared core self. It means to explore entanglement by getting entangled more deeply. Art is an exploration of the poetic space by means of embodied form. All art is sensual, all of its works are to be perceived in space and time through a body. The functioning of organism is part of that poetic space of imagination. This space is neither “inside” in the brain or body nor does it exist beyond everything material in a “mental world”, but is rather the meaning of the metabolic and perceptual entanglements of self/other.

Poetic space is a dimension that comes whith organism. It is that which Spinoza had named “Conatus”, the intermingled unfolding of living processes, experienced from the inside. Alive, we are in full “conativity”, and we have organs to experience and to enhance it as we have organs to destroy it. Conativity is the way reality unfolds for us living beings. It is the process into which the condition of life opens out. It happens in poetic space, as the continued unfolding of it. What relativity is for physics, conativity is for biology.

The poetic space is the area which is spanned out through entanglement. Nature is self-expressive, mutually transformative, hence inter-imaginative. Nature ist the arena of poetic space. The whole of nature is inter-imagination and interpenetration. Aliveness therefore is not the model of healthy salvation, aliveness is ongoing inter-imagination. This inter-imagination takes place as meaning, based on and always connected with, but not reducible to material or causal relationships.

Connectedness in Freedom

In this book I have unfolded the argument that we have to drop the seperation into inside and outside, so engrained in our culture, and, as we are formed by this culture, also in our everyday experience, and, most profoundly, in our science. I have tried to show that the mystery of organism lies in the fact that it actually forms a gate between these dimensions. In organism, outside always is inside, as every material impact has an existential meaning for the process of life, and every experienced (or imagined, or created) meaning has a material consequence, because the coherence of the individual is based on meaning, not on causal chains. This existential primacy is the realization of poetic space.

This space is poetic, namely meaningful, because it is about the meanings of relationships, and it is poetic, namely expressive, because these relationships can only be material transformations, and hence their meanings must necessarily be embodied, which is why they are significant about their underlying import. Meaning in matter is poetic expression. Because we can only relate to this meaning through

[1]translation by me, A.W. The original reads: »Kunst verhält sich zur Schöpfung gleichnisartig. Sie ist jeweils ein Beispiel, ähnlich wie das Irdische ein kosmisches Beispiel ist.«

our senses as an embodied being who itself is meaning in matter, every poetic expression is necessarily entanglement.

Poetic matter is always a part of the whole which shines through its parts. The creamy yellow of autumns leaves shows that the cosmos is capable of producing a quality of tender creaminess, and hence, that there is the existential dimension of smooth yet intense energy. The whole displays its possibilities through its parts. It yearns to be more deeply intensifyed and understood through its parts in what would ultimately be an understanding of itself. Michael McCarthy (2015:160) grasps this when he writes of the colour of the wings of the large copper butterfly, as "purest, most saturate orange you can possibly conceive of; indeed it may be more than you conceive of, which is perhaps the source of the delight when you eventually set eyes on it, as your sense of what the world can contain is suddenly enlarged." This is the full-fledged imaginary surplus Varela first spoke of.

But poetic expression does not only connect the specific embodied phenomenon to whole. It also entangles outside with inside, and hence generates new outside expressions from inside experiences. It has consequences outside its proper time and space, and even backwards in chronological time: Any intense emotional experience can make you re-evaluate your own existence in an instant.

To be entangled, to partake in wandering together through poetic space, creates delight. It is joy to enter into connection, to see insides revealed as effects of outsides, and to experience the increase in depth through the encounter with imagination. The poetic space is what enlivens us, as it is our true space, so much as it is the home of every other living thing. Being connected to the poetic space makes us feel joy. It makes us smile involuntarily, or laugh with glee, as kids do, or, expressed in the elastic dance of their limbs, young puppies and infant antelopes. We might even say that joy is the realization of entanglement-through-entanglement, and hence the epitome, but also the deep center, of being alive. Joy means to realize our connectedness in freedom.

This is what the German poet and philosopher Friedrich Schiller (2000) thought the experience of beauty was indicative of: For Schiller beauty – and joy – materialize if we are able to imagine the necessity of entanglement in an individual way. The experience of beauty is the perception of entanglement through entanglement. It means to perceive the other through the felt transformations of self. It makes us feel the paradoxical union of seperatedness and fusion in the center of life. Realizing entanglement means to realize oneself-in-connection, and through this to assert one's own autonomy, which is at the same time dependence and interpenetration. The infection with aliveness is the signifying character of this experience, as it so deeply tells about the character of our individuality in the world, and the relationship both continuously evolve together.

We can find the experience of self-in-connection also in relation to a work of art, where a single word can have the "reservoir of darkness stirred" (W. H. Auden) and make us understand ourselves better through realizing that which we find before us in a drawing, or hear as a phrase, or are cradled by through a tune and a rhythm, is actually a real part of our inner landscape. We find this experience in the face of every other creature. The other subject is to some extent also ourselves, cognitively tied to us through the workings of the mirror systems of the brain. We find it again in the tender gaze of another human meaning in her eyes' expression that she wants us to be.

Only in entanglement we can be autonomous. This is the insight still missing from technical explanations of the puzzle posed by quantum mechanics. This insight understands the impact of entanglement through the action of feeling. Only in incorporating life's subjectivity and the ensuing meaningful – poetic – entanglement, we can do justice to the physical picture from an embodied standpoint. What physics misses is not an even more complicated quantum mechanics of organism, but the simple fact that through the imaginary surplus which comes into being through the yearning bodies of feeling subjects, time and locality are cancelled and the observer is always changed by what he observes. There is no observation without feeling. Perception is entanglement, as is breathing and eating. The body and all the connections it constantly makes whispers to us that entanglement is the default way experience can come into being.

Feeling, the existential experience of meaning, is at the same time the basic experience of being interconnected and of being separate. Feeling is the experience of inside through an outside, the changes done to an outside, and the joy manifest in an outside, and hence an outside becoming pure inwardness, and still remaining matter. Feeling hence witnesses the crossing of the domains of matter and mind. It binds them together because it is in both of them, and, as life itself, is only possible through this junction, which is matter gaining an interest in itself, matter speculating about possible other fates to be had, possible other existential dramas, in its unquenchable thirst to become real, in its insatiable desire to feel itself through remembering what it is in all which it is not. And this, the undying interest of becoming self through totally giving self away to the other, is the only poetic principle and the sole prerequisite for tenderness that was ever needed.

Bibliography

Abram, D. (1997). *The spell of the sensuous. Perception and the more-than-human-world*. New York: Pantheon.

Adorno, T. W. (2013). *Aesthetic theory*. New York: Bloomsbury.

Albert, C. (1991). *Tagebücher 1935–1951*. Reinbek bei Hamburg: Rowohlt.

Aristotle. *De anima* II, 1, 412a.

Aristotle. *Physik* 254b.

Babiloni, F., & Astolfi, L. (2014). Social neuroscience and hyperscanning techniques: Past, present and future. *Neuroscience and Biobehavioral Reviews, 44*, 76–93.

Bachelard, G. (1942). *L'eau et les rèves. Essai sur l'imagination de la matière*. Paris: Corti.

Bachelard, G. (1990). *Psychoanalyse des Feuers* (A. W., Trans.). Frankfurt am Main: Fischer.

Bateson, G. (1972). *Steps to an ecology of mind*. New York: Ballantine.

Bateson, G. (1982). *Geist und Natur. Eine notwendige Einheit*. Frankf. am. Main: Suhrkamp.

Bateson, G. (1988). *Ökologie des Geistes*. Frankfurt am Main: Suhrkamp.

Bateson, G., & Bateson, M. C. (1993). *Wo Engel zögern. Unterwegs zu einer Epistemologie des Heiligen*. Frankfurt am Main: Suhrkamp.

Bateson, G., & Bateson, M. C. (2004). *Angels fear: Towards an epistemology of the sacred*. New York: Hampton Press.

Bauer, J. (2006). *Warum ich fühle, was du fühlst: Intuitive Kommunikation und das Geheimnis der Spiegelneurone*. München: Heyne.

Bauer, J. (2008). *Das kooperative Gen*. Hamburg: Hoffmann und Campe.

Bitbol, M. (2010). L'écologie en première personne. *Talk, CREA Paris*, 6. April 2010.

Blumenberg, H. (1981). *Die Lesbarkeit der Welt*. Frankfurt am Main: Suhrkamp.

Böhler, D. (1994). Hans Jonas. Stationen, Einsichten und Herausforderungen eines Denklebens. In D. Böhler (Ed.) *Ethik für die Zukunft. Im Diskurs mit Hans Jonas* (pp. 45–68). München: C.H. Beck.

Bohm, D. (1985). *Unfolding meaning: A weekend of dialogue with David Bohm*. New York: Routledge-Ark Paperbacks.

Bohm, D. (1990). A new theory of the relationship of mind and matter. *Philosophical Psychology, 3*(2), 271–286.

Böhme, H. (1988). *Natur und Subjekt*. Frankfurt am Main: Suhrkamp.

Böhme, G. (1992). *Natürlich Natur*. Frankfurt am Main: Suhrkamp.

Böhme, G., & Böhme, H. (1996). *Feuer, Wasser, Erde, Luft: Eine Kulturgeschichte der Elemente*. München: C. H. Beck.

Böhme, H., Müller, L., & Matussek, P. (2000). *Orientierung Kulturwissenschaft: Was sie kann, was sie will*. Reinbek bei Hamburg: Rowohlt.

A. Weber, *Biopoetics*, Biosemiotics 14, DOI 10.1007/978-94-024-0832-4

Bollier, D. (2014). *Think like a commoner. A short introduction to the life of the commons*. Gabriola Island: New Society Publishers.
Borges, J. L. (1992). Das Aleph. In Ders. *Das Aleph. Erzählungen 1944–1952*. Frankfurt am Main: Suhrkamp.
Brooks, D. R., & Wiley, E. O. (1988). *Evolution as entropy. Toward a unified theory of biology.*. Chicago/London: University of Chicago Press.
Bryant, J. (2006). To fight some other world. In B. Bryant & M. Edwards, (Eds.), *Ungraspable phantom: Essays on Moby Dick*. Kent: Kent State University Press.
Buber, M. (1937). *I and Thou*. Eastford: Martino (2010 Reprint of the original American edition).
Calvino, I. (1976). Blood, sea. In *t zero*. New York: Mariner Books.
Campbell-Fisher, I. G. (1950). Aesthetics and the logic of sense. *Journal of General Psychology, 43*, 245–273.
Carroll, S. B. (2005). *Endless forms most beautiful. The new science of Evo Devo and the making of the animal kingdom*. New York & London: Norton.
Carter, J. (2014). *Ricoeur on moral religion: A hermeneutics of ethical life*. Oxford: Oxford University Press.
Cassirer, E. (1982). *Philosophie der symbolischen Formen*. Darmstadt: Wissenschaftliche Buchgesellschaft.
Cassirer, E. (1994). *An essay on man*. New Haven und London: Yale University Press (Reprint 1992).
Cassirer, E. (1995). *Nachgelassene Manuskripte und Texte*. Herausgegeben von John Michael Krois und Oswald Schwemmer. In *Band 1: Zur Metaphysik der symbolischen Formen*. Hamburg: Meiner.
Cassirer, E. (2010). *Philosophie der Symbolischen Formen*. Stuttgart: Meiner.
Chalmers, D. (1996). *The conscious mind*. New York: Oxford University Press.
Chernyak, L., & Tauber, A. I. (1991). The dialectical self: Immunology's contribution. In A. I. Tauber (Ed.), *Organism and the origins of self*. Dordrecht: Kluwer.
Clark, A. (2008). *Supersizing the mind: Embodiment, action, and cognitive extension*. New York: Oxford University Press.
Coetzee, J. M. (1999). *The lives of animals*. Princeton: Princeton University Press.
Coetzee, J. M. (2004). *Elizabeth Costello*. Frankfurt am Main: Suhrkamp.
Conrad, J. (2000). *Lord Jim. Eine Geschichte*. München: Piper.
Cornell, J. F. (1986). A Newton of the grassblade? Darwin and the problem of organic teleology. *Isis, 77*, 405–421.
Corrington, R. S. (1995). *Ecstatic naturalism: Signs of the world*. Bloomington: Indiana University Press.
Damasio, A. (2000). *The feeling of what happens: Body and emotion in the making of consciousness*. New York: Harcourt Brace.
Darwin, C. R. (1859). *On the origin of species by means of natural selection, or the preservation of favoured races in the struggle for life*. London: John Murray.
De Jaegher, H. (2015). How we affect each other. Michel Henry's 'Pathos-With' and the enactive approach to intersubjectivity. *Journal of Consciousness Studies, 22*(1–2), 112–132.
De Jaegher, H., & Di Paolo, E. T. (2007). Participatory sense-making: An enactive approach to social cognition. *Phenomenology and the Cognitive Sciences, 6*, 485–507.
Deacon, T. (2012). *Incomplete nature. How mind emerged from matter*. New York: Norton.
Deleuze, G. (1988). *Le Pli*. Paris: Minuit.
Derrida, J. (1990). *Qu'est-ce qu'est la poésie?/Was ist Dichtung?/Che cos'è la poesia?/What is poetry?* (A. Garcia Düttmann, M. Ferraris, & P. Kamuf, Trans.). Berlin: Brinkmann & Bose.
Descola, P. (1997). *Par-delà nature et culture*. Paris: Gallimard.
Di Paolo, E. A. (2005). Autopoiesis, adaptivity, teleology, agency. *Phenomenology and the Cognitive Sciences, 4*, 97–125.
Di Paolo, E., & De Jaegher, H. (2016). Neither individualistic, nor interactionist. In C. Durt, T. Fuchs, & C. Tewes (Eds.), *Embodiment, enaction, and culture*. Cambridge: MIT Press.

Di Paolo, E., & Thompson, E. T. (2014). The enactive approach. In L. Shapiro (Ed.), *The routledge handbook of embodied cognition*. London: Routledge Press.
Dumas, G., Nadel, J., Soussignan, R., Martinerie, J., & Garnero, L. (2010). Inter-brain synchronization during social interaction. *PloS One, 5*, e12166.
Eccles, J. C. (1994). *How the self controls its brain*. Berlin/New York: Springer.
Edelman, G. M. (1988). *Topobiology. An introduction to molecular embryology.*. New York: Basic Books.
Eliade, M. (1990). *Das Heilige und das Profane*. Frankfurt am Main: Suhrkamp.
Emerson, R. W. (1951). The poet. In *Emerson's essays* (pp. 261–291). New York: Harper and Row.
Feyerabend, P. (2009). *Naturphilosophie*. Frankfurt am Main: Suhrkamp.
Foucault, M. (1990). Einleitung zu *Der Gebrauch der Lüste*. In P. Engelmann (Ed.), *Postmoderne und Dekonstruktion. Texte französischer Philosophen zur Gegenwart* (pp. 84–107). Stuttgart: Philipp Reclam jun.
Fox Keller, E. (1999). *Le rôle des métaphores dans le progrès de la biologie*. Le Plessis-Robinson: Institut Synthélabo.
Fox Keller, E. (2002). *Making sense of life. Explaining biological development with models, metaphors and machines.* Cambridge: Harvard University Press.
Frye, N. (1991). *Double vision. Identity and meaning in religion.* Toronto: Toronto University Press.
Gaità, R. (2005). *The philosopher's dog. Friendships with animals*. New York: Random House.
Gefter, A. (2007). Once more with feeling. Interview with Marvin Minsky. *The New Scientist, 193*(2592), 48–49.
Glacken, C. J. (1967). *Traces on the rhodian shore: Nature and culture in western thought from ancient times to the end of the eighteenth century.* Berkeley/Los Angeles: California University Press.
Glissant, É. (1996). *Introduction à une Poétique du Divers*. Paris: Gallimard.
Glissant, É. (1997). *Poetics of relation*. East Lansing: Michigan State University Press.
Glissant, É. (2002, April 19). The poetics of the world: Global thinking and unforeseeable events. *Chancellor's distinguished lecture,* Louisiana State University: Baton Rouge.
Glissant, É. (2005). Comme l´oiseau innumérable. In *La cohée du Lamentin*. Poétique V, Paris: Gallimard.
Goethe, J. W. (1998). *Gesammelte werke*. Hamburger Ausgabe. München: dTV
Gopnik, A. (2010). *Kleine Philosophen: Was wir von unseren Kindern über Liebe, Wahrheit und den Sinn des Lebens lernen können*. Berlin: Ullstein.
Gouin, J.-P., Carter, C. S., Pournajafi-Nazarloo, H., et al. (2010). Marital behavior, oxytocin, vasopressin, and wound healing. *Psychoneuroendocrinology, 35*, 1082–1090.
Hameroff, S., & Penrose, R. (2014). Consciousness in the universe A review of the 'Orch OR' theory. *Physics of Life Reviews, 11*(2014), 39–78.
Hamilton-Paterson, J. (1995). *Seestücke*. Stuttgart: Klett-Cotta.
Hardt, M., & Negri, A. (2009). *Commonwealth*. Harvard: Harvard University Press.
Harrington, A. (1997). *Reenchanted science. Holism in German philosophy and science from Wilhelm II. to Hitler.* Cambridge: Harvard University Press.
Hegel, G. F. W. (1969). *Science of logic* (M. J. Miller, Trans.). New York: Humanities Press.
Hegel, G. W. F. (1970). *Hegel's philosophy of nature* (M. J. Petry Trans.). London: Allen & Unwin.
Hegel, G. W. F. (1991). *The encyclopedia logic. Part I of the encyclopaedia of philosophical sciences with the Zusätze* (T. E. Geraets, W. A. Suchting, & H. S. Harris, Trans.). Indianapolis-Cambridge: Hackett.
Hoffmeyer, J. (1997). *Signs of meaning in the universe.* Bloomington: Indiana University Press.
Hoffmeyer, J. (2006). Genes, development and semiosis. In E. Neumann-Held & C. Rehmann-Sutter (Eds.), *Genes in development. Re-reading the molecular paradigm* (pp. 152–174). Durham/London: Duke University Press.
Hoffmeyer, J. (2014a). The semiome: From genetic to semiotic scaffolding. *Semiotica, 198*, 11–31.

Hoffmeyer, J. (2014b). Semiotic scaffolding: "A biosemiotic link between Sema and Soma". In K. R. Cabell & L. Valsiner (Eds.), *The catalyzing mind* (Annals of theoretical psychology, Vol. 11, pp. 95–110).

Hopkins, G. M. (2009). *Journal* (P. Waterhouse, Trans.). Salzburg: Jung & Jung.

Hyde, L. (2007). *The gift. Creativity and the artist in the modern world*. New York: Random House.

Igamberdiev, A. U. (1999). Semiosis and reflectivity in life and consciousness. *Semiotica, 123*(3/4), 231–246.

Ingold, T. (2000). *The perception of the environment: Essays on livelihood, dwelling and skill*. London: Routledge

Jablonka, E., & Lamb, M. (2005). *Evolution in four dimensions. Genetic, epigenetic, behavioral, and symbolic variation in the history of life*. Cambridge/London: MIT Press.

Jonas, H. (1966). *The phenomenon of life*. New York: Harper & Row.

Jonas, H. (1973). *Organismus und Freiheit: Ansätze zu einer philosophischen Biologie*. Göttingen: Vandenhoeck und Ruprecht.

Jonas, H. (1992): *Philosophische Untersuchungen und metaphysische Vermutungen*. Frankfurt am Main: Insel-Verlag.

Jonas, H. (1996). *Das Prinzip Leben*. Frankfurt am Main: Insel.

Juarero, A. (1999). *Dynamics in action. Internal behaviour as a complex system*. Cambridge: MIT Press.

Kauffman, S. (1996). *At home in the universe: The search for the laws of self-organization and complexity*. Oxford: Oxford University Press.

Kauffman, S. (2002). *Investigations*. Oxford: Oxford University Press.

Kauffman, S. (2016). *Humanity in a creative universe*. New York: Oxford University Press.

Kiecolt-Glaser, J. K., Loving, T. J., Stowell, J. R., et al. (2005). Hostile marital interactions, proinflammatory cytokine production, and wound healing. *Archives of General Psychiatry, 62*, 1377–1384.

Kirschner, M. W., & Gerhart, J. C. (2005). *The plausibility of life. Recolving Darwin's Dilemma*. New Haven: Yale University Press.

Kirschner, M. W., Gerhart, J. C., & Mitchison, T. (2000). Molecular 'Vitalism'. *Cell, 100*, 79–88.

Krampen, M. (1992). Phytosemiotics revisited. In T. A. Sebeok & J. Umiker-Sebeok (Eds.), *Biosemiotics: The semiotic web 1991* (pp. 213–219). Berlin: Mouton de Gruyter.

Krebs, A. (Ed.). (1997). *Naturethik: Grundtexte der gegenwärtigen tier- und ökoethischen Diskussion*. Frankfurt am Main: Suhrkamp.

Kulesa, P. M., & Bonner-Fraser, M. (2000). In Ovo time-lapse analysis after Dorsal Neural Tube Ablation shows rerouting of chick hindbrain neural crest. *Development, 127*(13), 2843–2852.

Kull, K. (1999). Biosemiotics in the twentieth century: A view from biology. *Semiotica, 127*(1/4), 385–414.

Kull, K. (2012). Introduction. In R. Silver & B. Tyler, (Eds.), *Gatherings in biosemiotics* (Tartu Semiotics Library 11). Tartu: University of Tartu Press.

Kull, K. (2015a). Semiosis stems from logical incompatibility in organic nature: Why biophysics does not see meaning, while biosemiotics does. *Progress in Biophysics and Molecular Biology, 119*, 616–621. http://dx.doi.org/10.1016/j.pbiomolbio.2015.08.002

Kull, K. (2015b). Evolution, choice, and scaffolding: Semiosis is changing its own building. *Biosemiotics*. doi:10.1007/s12304-015-9243-2

Kurt, H. (2011). *Wachsen. Über das Geistige in der Nachhaltigkeit*. Stuttgart: Meyer.

Kyselo, M. (2014). The body social: An enactive approach to the self. *Backiers in Psychology, 5*, 1–16.

Lachmann, R. (1999). Animal perception as value perception. *Evolution and Cognition (New Series), 5*, 189–198.

Lachmann, R. (2000). *Susanne K. Langer: Die lebendige Form menschlichen Fühlens und Verstehens*. München: Fink.

Lakoff, G., & Johnson, M. (1999). *Philosophy in the flesh*. New York: Basic Books.

Langer, S. K. (1953). *Feeling and form*. New York: Scribner's.

Langer, S. K. (1967–1983). *Mind: An essay on human feeling* (Vol. 3). Baltimore: Johns Hopkins University Press.
Langer, S. K. (1979). *Philosophie auf neuem Wege*. Mittenwald: Mäander Kunstverlag.
Latour, B. (1993). *We have never been modern*. Cambridge, MA: Harvard University Press.
Layton, R. (1995). Relating to the country in the Western Desert. In E. Hirsch & M. O'Hanlon (Eds.), *The anthropology of landscape* (pp. 210–231). Oxford: Clarendon Press.
Leopold, A. (1949). *A sand county almanach and sketches here an there*. New York/Oxford: Oxford University Press.
Lévi-Strauss, C. (1968). *Das wilde Denken*. Frankfurt am Main: Suhrkamp.
Ley, R. E., Lozupone, C. A., Hamady, M., Knight, R., & Gordon, J. I. (2008). Worlds within worlds: Evolution of the vertebrate gut microbiota. *Nature Reviews, 6*, 776–788.
Libet, B. (1985). Unconscious cerebral initiative and the role of conscious will in voluntary action. *The Beavioural and the Brain Sciences, 8*, 529–566.
Lispector, C. (2014). *Agua viva*. London/New York: Penguin.
Löwy, I. (1991). The immunological construction of the self. In A. I. Tauber (Ed.), *Organism and the origins of self* (pp. 43–75). Dordrecht: Kluwer.
Lubarsky, S. (2014). Living beauty. In G. Wuerthner, E. Crist, & T. Butler (Eds.), *Keeping the wild: Against the domestication of the earth* (pp. 194–195). Washington, DC: Island Press.
Macfarlane, R. (2007). *The wild places*. New York: Penguin Books.
Margulis, L. (1999). *Symbiotic planet: A new look at evolution*. New York: Basic Books.
Margulis, L., & Guerrero, R. (1991). Two plus three equal one. Individuals emerge from bacterial communities. In W. I. Thompson (Ed.), *Gaia 2. Emergence. The new science of becoming*. Hudson: Lindisfarne Press.
Maturana, H. R., & Varela, F. J. (1980). *Autopoiesis and cognition: The realization of the living*. Boston: Reidel.
Maturana, H. R., & Varela, F. J. (1984). *Der Baum der Erkenntnis*. München: Goldmann.
McCarthy, M. (2015). *The moth snowstorm. Nature and joy*. London: John Murray.
Meltzoff, A. N. (2005). Imitation and other minds: The 'Like-Me- Hypothesis', In S. Hurley & N. Chater (Eds.), *Perspectives on imitation: From neuroscience to social science* (Vol. 2, pp. 55–77), Cambridge, MA: Harvard University Press.
Meltzoff, A. N., & Moore, M. K. (1995). Infants' understanding of people and things: From body imitation to folk psychology. In J. Bermúdez, A. Marcel, & N. Eilan (Eds.), *The body and the self* (pp. 43–69). Cambridge: MIT Press.
Melville, H. (1994). *Moby Dick*. Zürich: Diogenes.
Merleau-Ponty, M. (1964). *Le visible et l'invisible*. Paris: Gallimard.
Merleau-Ponty, M. (1966). *Phänomenologie der Wahrnehmung*. Berlin: de Gruyter.
Michelini, F. (2012). Hegel's notion of natural purpose. *Studies in History and Philosophy of Biological and Biomedical Sciences, 43*, 133–139.
Midal, F. (2007). *Petit traité de la modernité dans l'art*. Paris: Pocket.
Midal, F. (2009). *Et si de l'amour on ne savait rien*. Paris: Albin Michel, impr.
Miller, H. (1981): "On Writing". In: *Reflections*. New York: Capra.
Miller, A. (1983). *Das Drama des begabten Kindes und die Suche nach dem wahren Selbst*. Frankfurt am Main: Suhrkamp.
Morin, E. (2001). *L'identité humaine. La methode, Tome 5, L'humanité de l'humanité*. Paris: Seuil.
Morphy, H. (1991). *Ancestral connections: Art and an aboriginal system of knowledge*. Chicago: University of Chicago Press.
Morphy, H. (1995). Landscape and the reproduction of the ancestral past. In E. Hirsch & M. O'Hanlon, (Eds.), *The anthropology of landscape* (pp. 184–209). Oxford: Clarendon Press.
Muir, J. (1979). *John of the mountains: The unpublished journals of John Muir.* (Reprinted by L. M. Wolfe, Ed., 1938, University of Wisconsin Press).
Müller, J. (2012). *Zur Vergleichenden Physiologie des Gesichtssinnes des Menschen und der Thiere*. South Carolina: Nabu Press.
Nagel, T. (1974). What is it like to be a Bat? *The Philosophical Review, 83*(4), 435–450.

Naumann, B., & Recki, B. (Eds.). (2002). *Cassirer und Goethe: neue Aspekte einer philosophisch-literarischen Wahlverwandschaft*. Berlin: Akademie-Verlag.

Nussbaum, M. (1986). *The fragility of goodness. Luck and ethics in Greek tragedy and philosophy*. Cambridge: Cambridge University Press.

Oelschlager, M. (1991). *The idea of wilderness: From prehistory to the age of ecology*. New Haven: Yale University Press.

Olbrich, E. (2009). Bausteine einer Mensch-Tier-Beziehung. In C. Otterstedt & M. Rosenberger (Eds.),*Gefährten – Konkurrenten – Verwandte. Die Mensch-Tier-Beziehung im wissenschaftlichen Diskurs*. Göttingen: Vandenhoeck und Ruprecht.

Oliver, M. (1986). *Dream Work*. New York: The Atlantic Monthly Press.

Oliver, M. (2004). Wild Geese. In *Wild Geese. Selected Poetry*. Tarset: Bloodaxe.

Orr, D. (2004). *Earth in mind: On education, environment, and the human prospect*. Washington, DC: Island Press.

Panksepp, J. (1998). The periconcious substrates of consciousness: Affective states and the evolutionary origins of self. *Journal of Consciousness Studies, 5*, 566–582.

Panksepp, J. (1999). Emotions as viewed by Psychoanalysis and Neuroscience: An exercise in consilience. *Neuropsychoanalysis: An Interdisciplinary Journal for Psychoanalysis and the Neurosciences, 1*, 15–38.

Panksepp, J. (2001). Affective neuroscience: Possible consilience between psychoanalysis and brain research. *Les états généraux de la psychanalyse*. http://www.etatsgeneraux-psychanalyse.net/archives/texte215.html

Panksepp, J. (2005). On the embodied nature of core emotional Affects. *Emotion Experience Journal of Consciousness Studies, 12*(8–10), 158–184.

Peil Kauffman, K. (2015). Emotional sentience and the nature of phenomenal experience. *Progress in Biophysics and Molecular Biology, 119*(3), 545–562. http://dx.doi.org/10.1016/j.pbiomolbio.2015.08.003

Peirce, C. S. (1968). *The collected papers of Charles Sanders Peirce*. Cambridge, MA: Harvard University Press.

Penrose, R. (1994). *Shadows of the mind. A search for the missing science of consciousness*. Oxford University Press: Oxford/New York/Melbourne.

Perry, S. (2003). *Coleridge's notebooks. A selection*. Oxford: Oxford University Press.

Plessner, H. (1928). *Die Stufen des Organischen und der Mensch*. Dritte, unveränderte Auflage. Berlin und New York: De Gruyter.

Plessner, H. (1982). Ein Newton des Grashalms? In Id., *Gesammelte Schriften* (pp. 247–266) Bd. 8, Frankfurt am Main: Suhrkamp.

Powell, D. (2009). Treat a female rat like a male and its brain changes. *New Scientist, 2690*, 8.

Reynolds, J. (2004). *Embodiment and the other. Relationships and alterity in phenomenology and deconstruction*. Ohio: Ohio University Press.

Ricoeur, P. (1992). *Oneself as Another* (*Soi-même comme un autre*) (K. Blamey, Trans.). Chicago: University of Chicago Press.

Rizzolatti, G., Fadiga, L., Gallese, V., & Fogassi, L. (1996). Premotor cortex and the recognition of motor actions. *Cognitive Brain Research, 3*(1996), 131–141.

Rodari, G. (1973). *Grammatica della Fantasia*. Torino: Einaudi.

Rosch, E. (1978). Principles of categorization. In E. Rosch & B. B. Lloyd (Eds.), *Cognition and categorization*. Hillsdale: Erlbaum.

Rosenberg, M. (2003). *Nonviolent communication: A language of life*. Encinitas: Puddle Dancer Press.

Roszak, T. (2000). Against single vision. In *Green studies reader. From romanticism to ecocriticism* (p. 11). New York/London: Routledge.

Ruyer, R. (1977). *La Gnose de Princeton. Des savants à la recherche d'une religion*. Paris: Hachette.

Schäfer, L. (1993). *Das Bacon-Projekt: Von der Erkenntnis, Nutzung und Schonung der Natur*. Frankfurt am Main: Suhrkamp.

Schama, S. (1996). *Der Traum von der Wildnis. Natur als Imagination*. München: Kindler.

Schrödinger, E. (1944). *What is life? The physical aspect of the living cell*. Cambridge: Cambridge University Press.
Schuster, P.-K. (2008). Die Welt als Fragment. Bausteine zm Universum Klee. In *Das Universum Klee*. Katalog zur Ausstellung. Hg. von Dieter Scholz und Christina Thomson. Ostfildern: Hatje Cantz Verlag, S. 15.
Schwemmer, O. (1997). *Ernst Cassirer. Ein Philosoph der Europäischen Moderne*. New York: De Gruyter.
Sebeok, T. A. (1963). Review of communication among social bees; porpoises and sonar; man and dolphin. *Language, 39*(3), 448–466.
Sewell, E. (1971). *The orphic voice. Poetry and natural history*. New York: Harper & Row.
Shephard, P. (1998). *Thinking animals. Animals and the development of human intelligence*. Athens/London: The University of Georgia Press.
Singh, N. M. (2013). The affective labor of growing forests and the becoming of environmental subjects: Rethinking environmentality in Odisha, India. *Geoforum, 47*, 189–198.
Snyder, G. (1992). *No nature: New and selected poems*. New York: Pantheon.
Snyder, G. (1990). *The practice of the wild*. Berkeley: Counterpoint.
Snyder, G. (2001). Language goes two ways. In C. McEwen, & M. Statman, (Eds.), *The alphabet of the trees. A guide to nature writing*. New York: Teachers & Writers Collaborative.
Steiner, G. (1991). *Réelles présences: Les arts du sens*. Paris: Gallimard.
Strohmann, R. (1997). The coming Kuhnian revolution in biology. *Nature Biot., 15*, 194–199.
Szymborska, W. (1996). *View with a grain of sand: Selected poems* (S. Baranczak & C. Cavanagh, Trans.). New York: Harcourt Brace.
Tefera, Z. (2015). The Guassa-Menz grazing commons. In S. Helfrich & D. Bollier, (Eds.), *Patterns of commoning*. Amherst: Levellers Press.
Thompson, E. T. (2007). *Mind in life: Biology, phenomenology, and the sciences of mind*. Cambridge, MA: Harvard University Press.
Van Horn, G. (2017a). *Channel coyotes*. Chicago: University of Chicago Press.
Van Horn, G. (2017b). Into the wildness. In G. Van Horn & J. Hausdoerffer (Eds.), *Wildness: Relations of people and place*. Chicago: Chicago University Press.
Varela, F. J. (1991). Organism: A meshwork of selfless selves. In A. I. Tauber (Ed.), *Organism and the origins of self*. Dordrecht: Kluwer Academic Publishers.
Varela, F. J. (1996). Neurophenomenology: A methodological remedy for the hard problem. *Journal of Consciousness Studies, 3*(4), 330–349.
Varela, F. J. (1997). Patterns of life: Intertwining identity and cognition. *Brain and Cognition, 34*, 72–87.
Varela, F. J. (2001). Intimate distances – Fragments for a phenomenology of organ transplantation. *Journal of Consciousness Studies, 8*(5–7), 259–271.
Varela, F. J., & Anspach, M. (1991). Immuknowledge: The process of somatic individuation, In W. I. Thompson (Ed.), *Gaia 2. Emergence, the new science of becoming* (pp. 68–85). Hudson: Lindisfarne, S.
Varela, F. J., Thompson, E. T., & Rosch, E. (1991). *The embodied mind: Cognitive science and human experience*. Cambridge: MIT Press.
Von Dassow, G., Meir, E., Munro, E. H., & Odell, G. M. (2000). The segment polarity network is a robust developmental module. *Nature, 406*, 188–192.
von Schiller, F. (2000). *Über die ästhetische Erziehung des Menschen: In einer Reihe von Briefen*. Hamburg: Reclam.
von Uexküll, J. (1980). *Kompositionslehre der Natur*. Frankfurt am Main und Berlin: Ullstein.
von Uexküll, T., Geigges, W., & Herrmann, J. M. (1993). Endosemiosis. *Semiotica, 96*(1/2), 5–51.
Weber, A. (1998). Der sprachlose Spiegel. *Deutsches Allgemeines Sonntagsblatt, 18*, 27–29.
Weber, A. (2001). Cognition as expression. On the autopoietic foundations of an aesthetic theory of nature. *Sign System Studies, 29*(1), 153–168.
Weber, A. (2002). Feeling the signs. The origins of meaning in the biological philosophy of Susanne K. Langer and Hans Jonas. *Sign Systems Studies, 30*(1), 183–200.

Weber, A. (2003). *Natur als Bedeutung. Versuch einer semiotischen Theorie des Lebendigen.* Würzburg: Königshausen & Neumann.
Weber, A. (2004). Mimesis and Metaphor: The biosemiotic generation of meaning in Cassirer and Uexküll. *Sign Systems Studies, 32.1*(2), 297–307.
Weber, A. (2007). *Alles fühlt. Mensch, Natur und die Revolution der Lebenswissenschaften*. Berlin: Berlin-Verlag.
Weber, A. (2010a). Der Seele tausend Masken. Warum wir ohne die Tiere uns selbst und die Welt nicht verstehen können. In *Wenn Tiere sich in der Theologie tummeln. Ansätze einer theologischen Zoologie.* Regensburg: Pustet.
Weber, A. (2010b). The book of desire: Towards a biological poetics. *Biosemiotics, 4*(2), 32–58.
Weber, A. (2012). There is no outside. A biological corollary for poetic space. In S. Rattasepp, & T. Bennett, (Eds.), *Gatherings in Biosemiotics* (Tartu Semiotics Library 11, pp. 225–226). Tartu: University of Tartu Press.
Weber, A. (2013). *Enlivenment. Towards a fundamental shift in the concepts of nature, culture and politics*. Berlin: Heinrich-Böll-Stiftung.
Weber, A. (2014). *Lebendigkeit. Eine erotische Ökologie*. München: Kösel.
Weber, A. (2015). Reality as commons. A poetics of participation for the anthropocene. In S. Helfrich & D. Bollier, (Eds.), *Patterns of commoning*. Amherst: Levellers Press.
Weber, A. (2016). *The biology of wonder. Aliveness, feeling, and the metamorphosis of science.* Gabriola Island: New Society Press.
Weber Nicholson, S. (2003). *The love of nature and the end of the world. The unspoken dimensions of environmental concern*. Cambridge, MA: The MIT Press.
Weber, B. H., & Depew, D. J. (1996). Natural selection and self-organization. Dynamical models as clues to a new evolutionary synthesis. *Biology and Philosophy*, 11, 33–65.
Weber, A., & Kurt, H. (2015). Towards cultures of aliveness. Politics and poetics in a postdualistic age. An anthropocene manifesto. *Solutions*, 5, 58–65.
Weber, A., & Varela, F. J. (2002). Life after Kant. Natural purposes and the autopoietic foundations of biological individuality. *Phenomenology and the Cognitive Sciences*, 1, 97–125.
Wetz, F. J. (1994). *Hans Jonas zur Einführung*. Hamburg: Junius.
Wilson, E. O. (1976). *Sociobiology*. Cambridge, MA: Belknap Press.
Wilson, E. O. (1984). *Biophilia*. Cambridge, MA: Harvard University Press.
Wilson, E. O. (1993). Biophilia and the conservation ethic. In E. O. Wilson, & S. E. Kellert (Eds.), *The biophilia hypothesis* (pp. 31–41). Washington, DC: Island Press, S.
Wilson, E. O. (1995). *Der Wert der Vielfalt: Die Bedrohung des Artenreichtums und das Überleben des Menschen*. München: Piper.
Wilson, E. O. (2012). *The social conquest of earth. A biological history of man*. Cambridge: Harvard University Press.
Wolfreys, J. (1998). *The Derrida reader: Writing performances*. Edinburgh: University of Edinburgh Press.
Zahavi, D. (2001). Beyond empathy. Phenomenological approaches to intersubjectivity. *Journal of Consciousness Studies, 8*(5–7), 151–167.
Zahavi, D. (2008). *Subjectivity and selfhood: Investigating the first-person perspective*. Cambridge, MA: MIT Press.

MIX
Papier aus verantwortungsvollen Quellen
Paper from responsible sources
FSC® C105338

If you have any concerns about our products,
you can contact us on
ProductSafety@springernature.com

In case Publisher is established outside the EU,
the EU authorized representative is:
Springer Nature Customer Service Center GmbH
Europaplatz 3, 69115 Heidelberg, Germany

Printed by Libri Plureos GmbH
in Hamburg, Germany